Gold Stars

KS2 MATHS & ENGLISH

Science too!

Ages 7-9

Written by
Paul Broadbent - Maths
Nina Filipek - English
Peter Riley - Science

PaRRagon

Bath · New York · Singapore · Hong Kong · Cologne · Delhi · Melbourne

Educational consultant:
Ann Dicks

Illustrated by
Rob Davis
and
Tom Connell

This edition published by Parragon in 2010

Parragon
Queen Street House
4 Queen Street
Bath BA1 1HE, UK

ISBN 978-1-4454-0364-9

Printed in China

Notes to parents

The Gold Stars® key stage 2 series

The Gold Stars® Key Stage 2 series has been created to help your child revise and practise key skills and information learned in school. Each book is a complete companion to the curriculum and has been written by an expert team of teachers.

How to use this book

- Do talk about what's on the page. Let your child know that you are sharing the activities. Talking about the sections that introduce and revise essential information is particularly important. Usually children will be able to do the fill-in activities fairly independently.

- Keep work times short. Do leave a page that seems too difficult and return to it later.

- It does not matter if your child does some of the pages out of turn.

- Your child may need some extra scrap paper for working out on some of the pages.

- Check your child's answers using the answer section on pages 172-183. Give lots of praise and encouragement and remember to reward effort as well as achievement.

- Do not become anxious if your child finds any of the pages too difficult. Children learn at different rates.

Contents

Maths

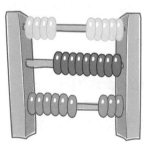

Contents

English

Contents

Science

Odd and even numbers

Learning objective: To recognise odd and even numbers.

Odd numbers always end in:

1	3	5	7	9

Even numbers always end in:

0	2	4	6	8

Divide by 2 to find out whether a number is odd or even.

13	21	16	20
* * * * * * * * *	* * * * * * * * * * *	* * * * * * * * *	* * * * * * * * * * *
* * * * * * * * *	* * * * * * * * * * *	* * * * * * * * *	* * * * * * * * * * *

Odd numbers always have 1 left over.
Even numbers can be put into 2 equal groups.

A

Join each of these numbers with a line to the correct box.

24   17   33
  15   29 22   38   30
     32   16   21
  20

ODD   EVEN

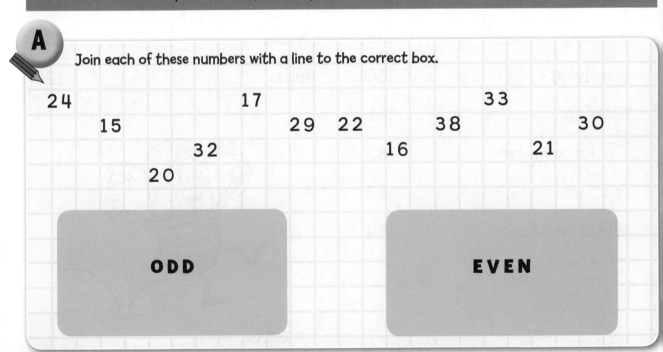

B

Circle all the odd numbers and underline all the even numbers.

①2③4⑤6⑦8 9 10 11 12 13 14 15 16 17 18 19 20

1. What is the next odd number after 10? ____

2. What is the next even number after 10? ____

3. What is the next odd number after 15? ____

4. What is the next even number after 14? ____

5. What is the next odd number after 20? ____

No matter how big they are, every single number except 0 is either odd or even.
Is 234,456,789 odd or even?

C

Write the next two numbers in these sequences.

1. 28 30 32 34 ___ ___

2. 19 21 23 25 ___ ___

3. 44 46 48 50 ___ ___

4. 25 27 29 31 ___ ___

Place value

Learning objective: To learn how 3-digit numbers are made.

3-digit numbers are made from hundreds, tens and ones.

Look at this number and how it is made:

593

five hundred and ninety-three

593 =
500 + 90 + 3

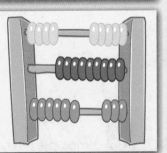

> The position of a digit in a number is really important. 359 and 593 use the same digits but are different numbers. The position of the digits 0 to 9 gives the value of the number.

A Write how many hundreds, tens and ones there are in each of these 3-digit numbers.

1. 398 = _____ + _____ + _____

2. 217 = _____ + _____ + _____

3. 452 = _____ + _____ + _____

4. 683 = _____ + _____ + _____

5. 165 = _____ + _____ + _____

6. 709 = _____ + _____ + _____

B

Write the missing numbers or words to complete each of these.

1. 9 4 1   →   nine hundred and _____

2. _____   →   three hundred and twenty-six

3. 5 3 4   →   _____

4. 8 7 0   →   _____

5. _____   →   two hundred and nineteen

6. _____   →   six hundred and fifty

C

Write the numbers shown on each abacus. The first one is done for you.

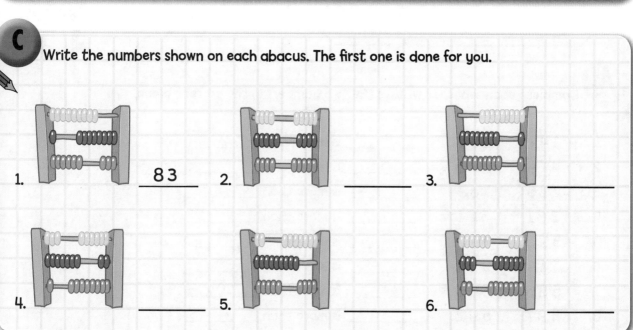

1. _83_   2. _____   3. _____

4. _____   5. _____   6. _____

Ordering numbers

Learning objective: To compare 3-digit numbers and put them in order.

$>$ is the sign for 'is more than'.
$<$ is the sign for 'is less than'.
$=$ is the sign for 'is equal to'.

198 $>$ 168

198 is more than 168.

126 $<$ 211

126 is less than 211.

You can remember which sign is which by looking at the shape. When we write 198 $>$ 168, then the sign is wider next to the larger number and narrower next to the smaller number.

You have to compare each digit in numbers to order them.

A Complete each sentence writing the two numbers in the correct place.

1.	147	152	_____ is less than _____ .
2.	479	476	_____ is less than _____ .
3.	735	753	_____ is more than _____ .
4.	381	521	_____ is more than _____ .
5.	390	190	_____ is less than _____ .
6.	214	244	_____ is less than _____ .
7.	586	585	_____ is more than _____ .
8.	497	592	_____ is more than _____ .

B

Write in the missing < or > signs for each pair of numbers.

1. 264 _____ 254

2. 328 _____ 431

3. 190 _____ 119

4. 536 _____ 523

5. 708 _____ 807

6. 655 _____ 635

C

Write each group of numbers in order starting with the smallest.

1. 159   191   112   125   _____ _____ _____ _____

2. 373   387   278   483   _____ _____ _____ _____

3. 645   668   622   739   _____ _____ _____ _____

4. 461   416   460   410   _____ _____ _____ _____

5. 743   778   760   704   _____ _____ _____ _____

6. 815   309   459   195   _____ _____ _____ _____

15

Number sequences

Learning objective: To continue number sequences by counting on or back in steps.

A number sequence is a list of numbers in a pattern. To find the rule or pattern in a sequence try finding the difference between each number.

Follow the rule to continue the sequence.

45 → 56 → 67 → 78 → 89
+11   +11   +11   +11

The rule or pattern is +11.

990 → 980 → 970 → 960 → 950
-10   -10   -10   -10

The rule or pattern is -10.

A

Write the next two numbers in each sequence.

1. 22   27   32   37   ____   ____

2. 85   87   89   91   ____   ____

3. 55   52   49   46   ____   ____

4. 44   48   52   56   ____   ____

5. 100   96   92   88   ____   ____

6. 519   509   499   489   ____   ____

7. 268   368   468   568   ____   ____

8. 931   831   731   631   ____   ____

If you find these activities easy peasy, stretch your brain with the next tricky challenges!

DEFINITION

number sequence A list of numbers in a pattern.

B

Write the missing numbers and the rule for each pattern.

#							
1.	70	____	100	115	130	____	The rule is : _____
2.	650	600	550	____	____	400	The rule is : _____
3.	____	843	____	643	543	443	The rule is : _____
4.	719	____	519	____	319	219	The rule is : _____
5.	____	825	830	835	____	845	The rule is : _____
6.	462	____	482	492	____	512	The rule is : _____
7.	133	123	____	____	93	83	The rule is : _____
8.	____	515	518	521	524	____	The rule is : _____

C

Two numbers in each sequence have been swapped over. Write each correct sequence.

1. 975   985   995   965   955   945   935

____ ____ ____ ____ ____ ____ ____

2. 80   180   280   380   680   580   480

____ ____ ____ ____ ____ ____ ____

3. 853   855   857   851   849   847   845

____ ____ ____ ____ ____ ____ ____

3-D solids

Learning objective: To name and describe 3-D solids.

There are 3-D solids all around you.

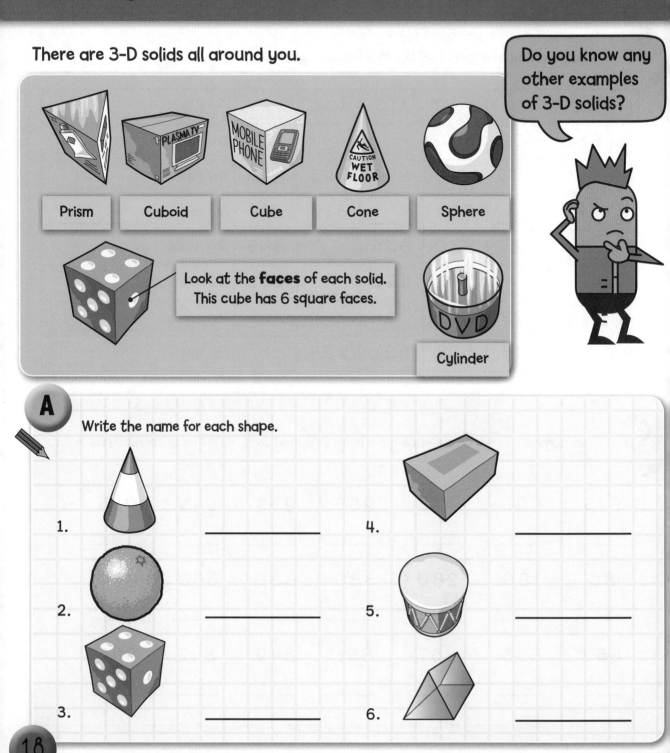

Do you know any other examples of 3-D solids?

| Prism | Cuboid | Cube | Cone | Sphere |

Look at the **faces** of each solid. This cube has 6 square faces.

Cylinder

A

Write the name for each shape.

1. _____

2. _____

3. _____

4. _____

5. _____

6. _____

DEFINITION

face The flat surface of a solid shape is called a face.

B

Name the shapes in each set and find the odd one out.

1. These shapes are all _____.
   The odd one out is a _____.

2. These shapes are all _____.
   The odd one out is a _____.

3. These shapes are all _____.
   The odd one out is a _____.

C

Complete this chart.

Name of shape	cube	cuboid	prism
Total number of faces			
Number of square and rectangle faces			
Number of triangular faces			

D

Are these statements **always, sometimes** or **never** true?

1. A cuboid has a triangular face. _____
2. A cone has a circle face. _____
3. A cylinder has two circle faces of different sizes. _____
4. A prism has a square face.
   _____

Number trios

Learning objective: To know addition and subtraction facts.

If you know an addition fact, you can work out a related subtraction fact.

Learning number trios is really useful.

Use trios of numbers, such as 11, 4 and 7 to learn the facts.

(11)(4)(7)

$4 + 7 = 11$ $11 - 4 = 7$

$7 + 4 = 11$ $11 - 7 = 4$

Use addition and subtraction facts to help with larger numbers.

$6 + 3 = 9$ $9 - 3 = 6$
$60 + 30 = 90$ $90 - 30 = 60$
$600 + 300 = 900$ $900 - 300 = 600$

A   Write the addition and subtraction families for each trio.

1.
| 7 | 8 | 15 |

___ + ___ = ___   ___ − ___ = ___
___ + ___ = ___   ___ − ___ = ___

2.
| 6 | 12 | 6 |

___ + ___ = ___   ___ − ___ = ___
___ + ___ = ___   ___ − ___ = ___

3.
| 9 | 14 | 5 |

___ + ___ = ___   ___ − ___ = ___
___ + ___ = ___   ___ − ___ = ___

4.
| 9 | 7 | 16 |

___ + ___ = ___   ___ − ___ = ___
___ + ___ = ___   ___ − ___ = ___

B

Answer these.

1. 6 + 9 = _____
   60 + 90 = _____
   600 + 900 = _____

2. 8 - 4 = _____
   80 - 40 = _____
   800 - 400 = _____

3. 7 + 5 = _____
   70 + 50 = _____
   700 + 500 = _____

4. 9 - 7 = _____
   90 - 70 = _____
   900 - 700 = _____

C

Write the missing numbers.

1. 6 + ☐ = 15

2. 13 − ☐ = 5

3. ☐ − 9 = 9

4. ☐ + 8 = 12

5. 40 + ☐ = 100

6. ☐ − 200 = 700

7. 80 − ☐ = 30

8. ☐ + 400 = 600

D

Work these out in your head.

1. What is the sum of 50 and 40?

_____

2. What is the total of 6 and 8?

_____

3. What is the difference between 14 and 7?

_____

4. Which number is 300 less than 900?

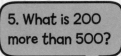

_____

5. What is 200 more than 500?

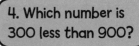

_____

6. What is 80 subtract 40?

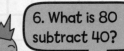

_____

21

Rounding numbers

We round numbers to make them easier to work with.
It is useful for estimating approximate, or rough, answers.

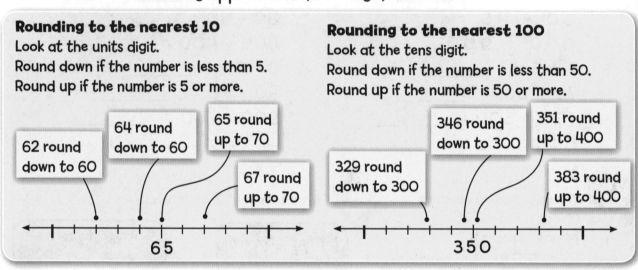

Rounding to the nearest 10
Look at the units digit.
Round down if the number is less than 5.
Round up if the number is 5 or more.

62 round down to 60

64 round down to 60

65 round up to 70

67 round up to 70

6 5

Rounding to the nearest 100
Look at the tens digit.
Round down if the number is less than 50.
Round up if the number is 50 or more.

329 round down to 300

346 round down to 300

351 round up to 400

383 round up to 400

3 5 0

A

Round these to the nearest 10.

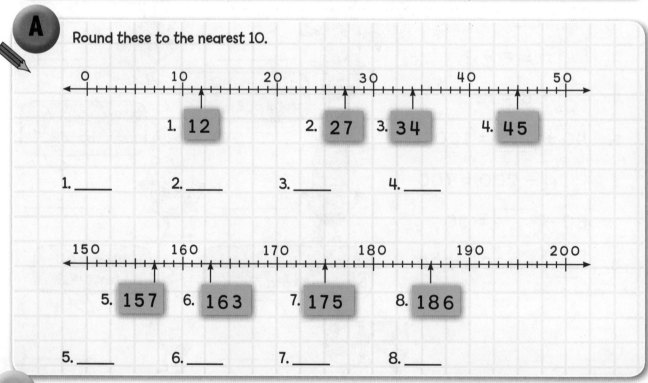

0   10   20   30   40   50

1. **12**   2. **27**   3. **34**   4. **45**

1. ____   2. ____   3. ____   4. ____

150   160   170   180   190   200

5. **157**   6. **163**   7. **175**   8. **186**

5. ____   6. ____   7. ____   8. ____

B

Round these to the nearest 100.

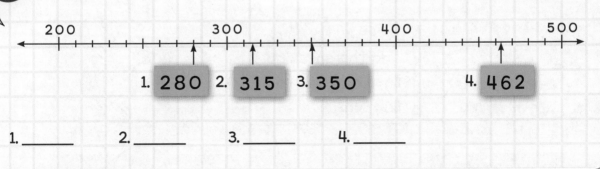

1. 280 2. 315 3. 350 4. 462

1. _____   2. _____   3. _____   4. _____

We can round numbers to find approximate answers.

Round to the nearest 10	Round to the nearest 100
24 + 67	481 - 176
20 + 70 = 90	500 - 200 = 300

C

Round to the nearest 10 and find an approximate answer.

1. 35 + 28

____ + ____ = ____

2. 97 – 51

____ – ____ = ____

3. 82 – 44

____ – ____ = ____

4. 146 + 43

____ + ____ = ____

Round to the nearest 100 and find an approximate answer.

5. 520 + 175

____ + ____ = ____

6. 842 – 305

____ – ____ = ____

7. 651 + 235

____ + ____ = ____

8. 793 – 619

____ – ____ = ____

Mental addition

Learning objective: To mentally add 1- and 2-digit numbers.

Break numbers up so that you can add them in your head.

What is 34 add 5?

34 + 5 =
30 + 4 + 5 =
30 + 4 + 5 = 39

Add the ones and then add this to the tens.

Add together 23 and 40.

23 + 40 =
20 + 3 + 40 =
20 + 40 + 3 = 63

Add the tens and then add on the ones.

A

Add the ones then add the tens and write the answer.

1. 63 + 4 =

2. 51 + 6 =

3. 32 + 3 =

4. 22 + 7 =

5. 94 + 4 =

6. 73 + 5 =

B

Add the tens then the ones and write the answer.

1. 56 + 20 =

2. 38 + 40 =

3. 52 + 30 =

4. 23 + 20 =

5. 11 + 70 =

6. 29 + 60 =

mental addition
Adding numbers in your head.

C

Join the pairs of sums with the same total.

55 + 20 = ☐   29 + 50 = ☐   62 + 6 = ☐

72 + 3 = ☐   71 + 8 = ☐   48 + 20 = ☐

37 + 30 = ☐   39 + 40 = ☐   74 + 5 = ☐   63 + 4 = ☐

D

Read the first statement then work out the questions in your head.

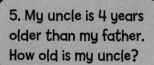

My mother is 34.

1. My aunt is 3 years older than my mother. How old is my aunt?

2. My father is 5 years older than my aunt. How old is my father?

3. My grandmother was 20 when my mother was born. How old is my grandmother?

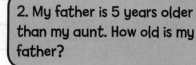

_____   _____    _____

4. My grandfather is 30 years older than my mother. How old is my grandfather?

5. My uncle is 4 years older than my father. How old is my uncle?

6. My great-grandmother is 50 years older than my mother. How old is my great-grandmother?

_____   _____   _____

Mental subtraction

Learning objective: To mentally subtract 1- and 2-digit numbers.

Break up numbers so that you can subtract them in your head.

What is 37 subtract 5?

$37 - 5 =$

$30 + 7 - 5 =$

$30 + 7 - 5 = 32$

Subtract the ones and then add this to the tens.

Take away 30 from 54.

$54 - 30 =$

$50 + 4 - 30 =$

$50 - 30 = 20 + 4 = 24$

Subtract the tens and then add on the ones.

A

Break up these numbers to subtract them in your head.

1. $46 - 4 =$ _____

2. $87 - 3 =$ _____

3. $29 - 6 =$ _____

4. $78 - 50 =$ _____

5. $93 - 30 =$ _____

6. $52 - 40 =$ _____

7. $72 - 30 =$ _____

8. $65 - 3 =$ _____

DEFINITION

find the difference
Another way of saying
'subtract' or 'take away'.

B

Find the difference between each pair of numbers.

1. | 98 | | 3 | _____
2. | 5 | | 57 | _____
3. | 3 | | 76 | _____

4. | 82 | | 40 | _____
5. | 20 | | 44 | _____
6. | 50 | | 91 | _____

C

Complete each chart to show the numbers coming out of each subtraction machine.

1.

IN 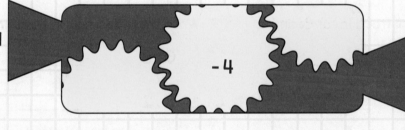 OUT

-4

IN	56	78	27	49	15	64
OUT	52					

2.

IN 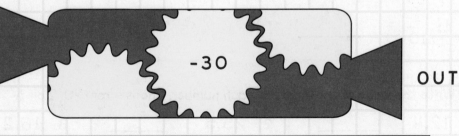 OUT

-30

IN	65	91	42	77	59	83
OUT	35					

27

Decimals

Learning objective: To use and understand tenths.

A decimal point is used to separate whole numbers from fractions. The digit after the decimal point shows the number of tenths.

Tenths break up a whole number into 10 equal parts.

$$\frac{1}{10} = 0.1 \qquad \frac{2}{10} = 0.2 \qquad \frac{3}{10} = 0.3$$

Example

15.7

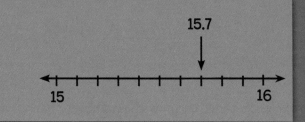

Tens   Units   Tenths

$$10 \ + \ 5 \ + \ \frac{7}{10} \ = \ 15.7$$

A

Write these fractions as decimals.

1. $6 \frac{3}{10}$ _____

2. $\frac{9}{10}$ _____

3. $12 \frac{4}{10}$ _____

4. $18 \frac{5}{10}$ _____

5. $11 \frac{1}{10}$ _____

Write these decimals as fractions.

1. 0.8 _____

2. 7.2 _____

3. 16.7 _____

4. 20.6 _____

5. 4.9 _____

B

Write the value of the digit 2 in each number. Choose from 20, 2 or $\frac{2}{10}$.

1. 12.4 _____   2. 25.5 _____   3. 16.2 _____

4. 3.2 _____   5. 0.2 _____   6. 42.1 _____

28

C

Look at these number lines and write the decimal number above each arrow.

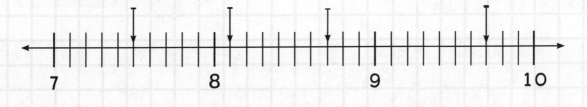

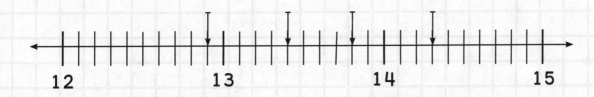

D

Write ⟨ or ⟩ between these numbers. Use the number lines above to help.

1. 7.6 _____ 7.3

2. 8.9 _____ 9.8

3. 13.4 _____ 12.4

4. 7.5 _____ 9.1

5. $9\frac{2}{10}$ _____ $7\frac{2}{10}$

6. $13\frac{9}{10}$ _____ $13\frac{8}{10}$

7. $8\frac{1}{10}$ _____ $9\frac{3}{10}$

8. $12\frac{4}{10}$ _____ $14\frac{2}{10}$

Look back at page 14 to find out what ⟨ and ⟩ mean.

2-D shapes

Learning objective: To name, draw and describe 2-D shapes.

2-D shapes are flat shapes. They can have straight or curved sides.

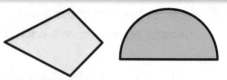

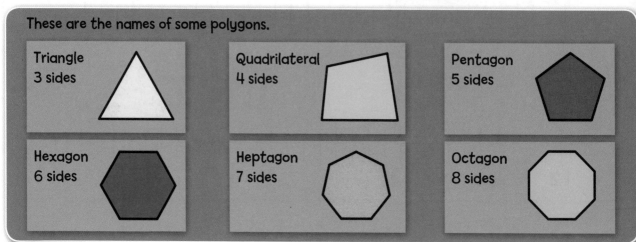

These are the names of some polygons.

Triangle
3 sides

Quadrilateral
4 sides

Pentagon
5 sides

Hexagon
6 sides

Heptagon
7 sides

Octagon
8 sides

A Write the name for each shape. Count the number of sides to help find the shape name.

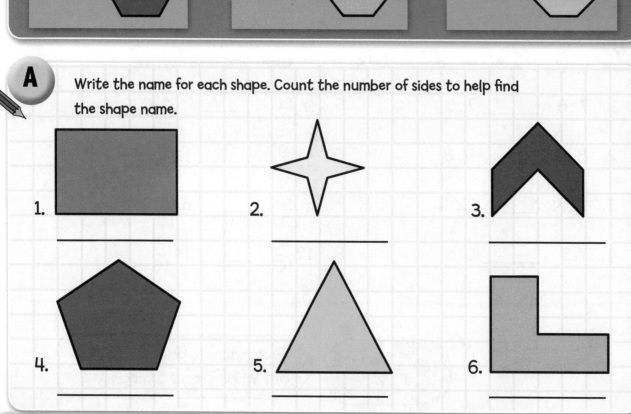

1. _____

2. _____

3. _____

4. _____

5. _____

6. _____

B

Write the name of the shapes in each set and the odd one out.

1. These shapes are _____. The odd one out is a _____.

2. These shapes are _____. The odd one out is a _____.

3. These shapes are _____. The odd one out is a _____.

C

A pentomino is made from joining five squares. Here are two examples.

Make other pentominoes from five squares. How many can you find?

What shapes are they?

31

Equivalent fractions

Learning objective: To use diagrams to identify equivalent fractions.

Some fractions are equivalent. This means they look different but have the same value.

These are all equivalent to $\frac{1}{2}$:

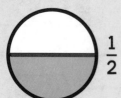 $\frac{1}{2}$  $\frac{2}{4}$  $\frac{3}{6}$  $\frac{4}{8}$

A Circle the rectangle that shows an equivalent fraction to each of the following numbers, then fill in the answer.

1. $\dfrac{1}{2}$ (a) (b) (c) (d) $\dfrac{1}{2} = \boxed{\dfrac{\quad}{\quad}}$

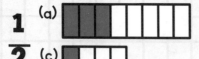

2. $\dfrac{1}{4}$ (a) (b) (c) (d) $\dfrac{1}{4} = \boxed{\dfrac{\quad}{\quad}}$

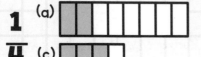

3. $\dfrac{1}{3}$ (a) (b) (c) (d) $\dfrac{1}{3} = \boxed{\dfrac{\quad}{\quad}}$

4. $\dfrac{1}{5}$ (a) (b) (c) (d) $\dfrac{1}{5} = \boxed{\dfrac{\quad}{\quad}}$

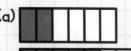

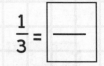
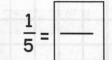

B

Look at the fractions that are shaded.
Write each fraction in two ways.

1. ⬚/⬚ = ⬚/⬚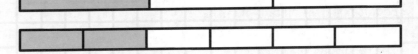

2. ⬚/⬚ = ⬚/⬚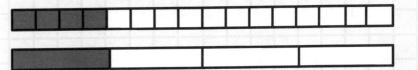

3. ⬚/⬚ = ⬚/⬚

C

Here are four members of the half family.

$$\frac{4}{8} \qquad \frac{6}{12} \qquad \frac{7}{14} \qquad \frac{10}{20}$$

Write five members of each of these fraction families.

1. $\frac{1}{3}$ ⬚   ⬚   ⬚   ⬚   ⬚

2. $\frac{1}{4}$ ⬚   ⬚   ⬚   ⬚   ⬚

3. $\frac{1}{5}$ ⬚   ⬚   ⬚   ⬚   ⬚

Measuring length

Learning objective: To read, estimate, measure and record using centimetres.

A ruler is a useful tool for measuring smaller lengths.

This shows a **centimetre** ruler.

- Each division is 1 centimetre in length.
- Each small division between the centimetres is half a centimetre.
- The length of the stick is 6 centimetres, or 6cm.

A Look at the ruler above and estimate the length of each line. Write your estimate in centimetres.

1. estimate: _____cm

2. estimate: _____cm

3. estimate: _____cm

4. estimate: _____cm

5. estimate: _____cm

6. estimate: _____cm

DEFINITION

estimate An estimate is a rough answer, without measuring.

Take your best guess!

34

B

Use a ruler and measure the exact length of each line in Section A. Write each length in centimetres.

1. length: _____cm

2. length: _____cm

3. length: _____cm

4. length: _____cm

5. length: _____cm

6. length: _____cm

C

Measure each item and write the lengths.

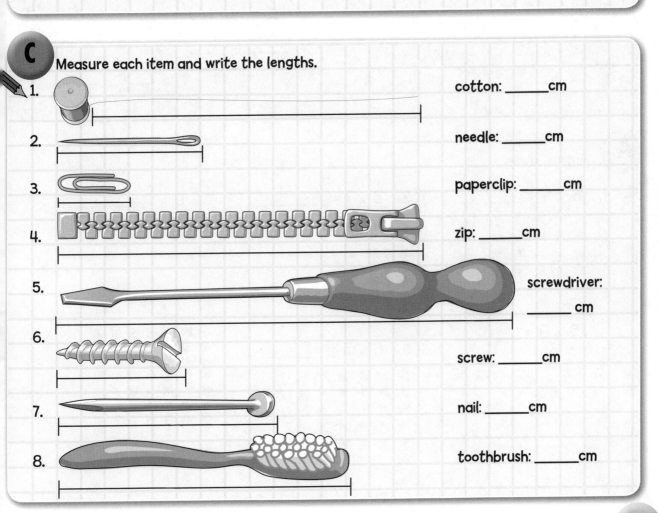

1.   cotton: _____cm

2.   needle: _____cm

3.   paperclip: _____cm

4.   zip: _____cm

5.   screwdriver: _____cm

6.   screw: _____cm

7.   nail: _____cm

8.   toothbrush: _____cm

Multiplication facts

Learning objective: To know the multiplication facts up to 10 x 10.

Use the multiplication facts you already know to help learn other facts.

Example

3 x 5 = 15

3 x 6 is 3 more ⟶ 18

8 x 2 = 16

8 x 4 is double 16 ⟶ 32

10 x 6 = 60

9 x 6 is 6 less ⟶ 54

Remember 3 x 7 gives the same answer as 7 x 3.

It's easy when you know the facts!

A   Answer these.

1.   6 x 4 = _____

2.   3 x 7 = _____

3.   9 x 3 = _____

6.   7 x 8 = _____

4.   5 x 9 = _____

5.   6 x 5 = _____

DEFINITION

multiplication fact
It is true that 3 x 5 = 15.
This is a multiplication fact.

B

Write the answers for each of these.

1. 5 x 7 = _____
   6 x 7 = _____

2. 5 x 8 = _____
   6 x 8 = _____

3. 10 x 6 = _____
   9 x 6 = _____

4. 10 x 8 = _____
   9 x 8 = _____

5. 3 x 3 = _____
   6 x 3 = _____

6. 2 x 7 = _____
   4 x 7 = _____

7. 4 x 4 = _____
   8 x 4 = _____

8. 2 x 9 = _____
   4 x 9 = _____

C

Read and answer these.

1. Julie buys 4 packs of plates. What is the total number of plates she will have? _____

2. Sam buys 3 packs of straws. How many straws does he have in total? _____

3. Owen wants 12 glasses. How many packs of glasses will he need to buy? _____

4. Owen also wants 12 plates. How many packs of plates will he need to buy? _____

5. Sabina wants 20 glasses. How many packs of glasses will she need to buy? _____

6. Tess wants 20 straws. How many packs of straws does she need to buy? _____

Now see if you can complete this tricky challenge!

2 x ☐ → 18
x   x
☐ x ☐ → 54
↓   ↓
12   81

☐ x 7 → 28
x   x
☐ x ☐ → 27
↓   ↓
36   21

37

Time

There are 60 minutes in 1 hour. It takes 5 minutes for the minute hand to move from one marker to the next.

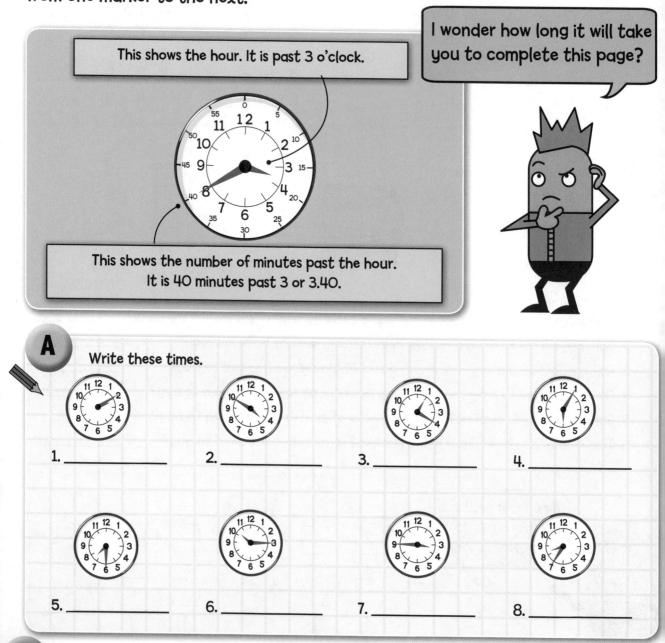

This shows the hour. It is past 3 o'clock.

I wonder how long it will take you to complete this page?

This shows the number of minutes past the hour.
It is 40 minutes past 3 or 3.40.

A Write these times.

1. _____

2. _____

3. _____

4. _____

5. _____

6. _____

7. _____

8. _____

B

Read these time problems. Write the answers.

1. A TV programme starts at 6.15 and lasts for half an hour.

   What time will it end?   _____

2. Nathan gets up at 7 o'clock and leaves for school an hour later.

   What time does he leave for school?   _____

3. A boat leaves at 10 past 1 and returns at half-past one.

   How long was the boat at sea?   _____

4. A cake takes 25 minutes to bake. It was put in the oven at 4 o'clock.

   When will it be ready?   _____

5. Gemma is playing at the park. It is quarter to 12. She has to go home at 12.30.

   How much longer does she have to play?   _____

C

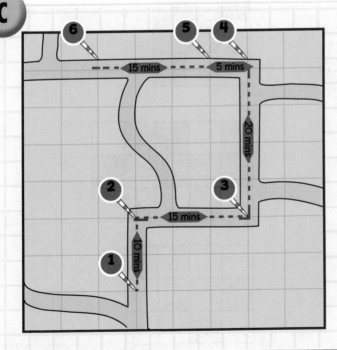

Look at the map showing the length of time a bus takes between each stop. Complete this bus timetable. Work out the time the bus will be at each stop.

Bus Stop	Time
1	9.05
2	
3	
4	
5	
6	

Measuring area

Learning objective: To find the area of shapes on a square grid.

To find the area of a shape you can draw it on a square grid and count the squares.

The side of each small square on this grid is 1cm.

This shape has an area of 12 square centimetres.

A Count the squares and write the area of each shape.

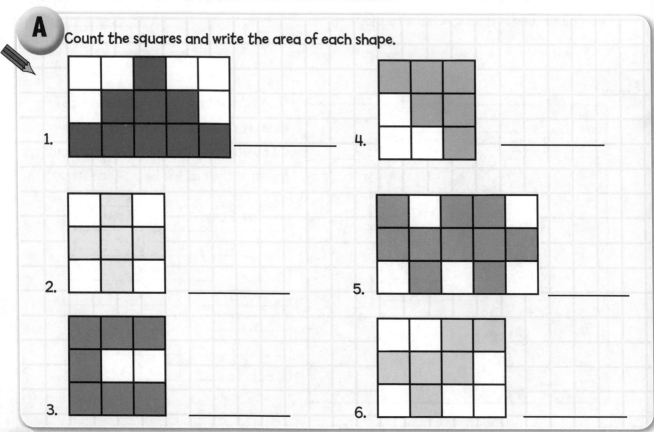

1. _____

2. _____

3. _____

4. _____

5. _____

6. _____

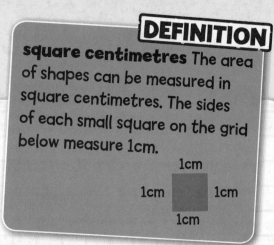

DEFINITION

square centimetres The area of shapes can be measured in square centimetres. The sides of each small square on the grid below measure 1cm.

B

This plan shows the gardens of a hotel. Each square shows 1 square metre of ground. Count the squares and write the area for each section.

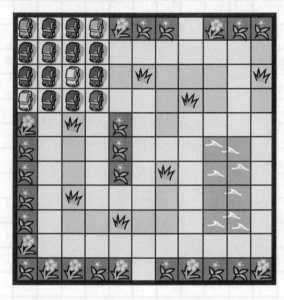

1. Area of swimming pool

= _____ square metres

2. Area of paths = _____ square metres

3. Area of car park = _____ square metres

4. Area of grass = _____ square metres

5. Area of flower border

= _____ square metres

C

A gardener is planning a path using 8 slabs. Each slab is 1 square metre. Here are two designs using 8 squares. Draw 3 more path designs using 8 squares.

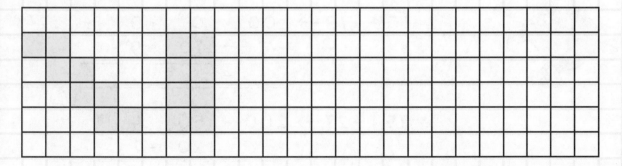

41

Written addition

Learning objective: To add 2- and 3-digit numbers.

When you can't work out an addition in your head, try a written method.

Example 1

46 + 25 → 40 + 6 + 20 + 5 = 60 + 11 = 71

Example 2

$$138 + 54$$

100 + 30 + 8
+   50 + 4
100 + 80 + 12 → 100 + 80 + 12 = 192

With this short method, add the ones, tens and then the hundreds.

```
  138
+  54
  192
    1
```

Make sure you line up the digits correctly.

A

Add these and write the answers.

1. 267 +18 → 200 + 60 + 7
           +   10 + 8
           _____
              +   +   = _____

2. 109 +79 → 100 + 0   + 9
           +   70 + 9
           _____
              +   +   = _____

3. 254 +27 → 200 + 50 + 4
           +   20 + 7
           _____
              +   +   = _____

B

Now answer these. Use paper for your working out.

1. 143
   + 37

2. 215
   + 29

3. 238
   + 58

4. 126
   + 35

C

Read and answer these. Use paper for your working out.

1. What is 15 more than 78? _____

2. Add 57 and 26. _____

3. What is the total of 33 and 39? _____

4. Increase 124 by 47. _____

5. Total 265 and 29. _____

6. What is 46 added to 205? _____

D

Read and answer these problems. Use paper for your working out.

1. A truck driver travels 53 kilometres in the morning and 37 kilometres in the afternoon.
   How far does the truck travel in total? _____

2. A market stall sells 28 bottles of mango juice and 39 bottles of orange juice.
   What is the total number of bottles sold? _____

3. A farmer has 44 chickens and 17 ducks.
   How many chickens and ducks are there altogether? _____

4. A postman has 149 letters and 36 parcels.
   How many items altogether are there to deliver? _____

5. Jamal has read 108 pages of his reading book and there are 52 pages left.
   How many pages in total are there in Jamal's reading book? _____

6. Julie is 136 centimetres tall and her dad is 38 centimetres taller than she is.
   How tall is Julie's dad? _____

Written subtraction

Learning objective: To subtract 2-digit numbers.

When you can't work out a subtraction in your head, try a written method.

Look at these two methods.

53 - 38

Break up 53 into 40 and 13:

$$\begin{array}{r} 4\ \ 13 \\ \cancel{5}\ \ \cancel{3} \\ -\ 3\ \ 8 \\ \hline 1\ \ 5 \end{array}$$

13 - 8 = 5   40 - 30 = 10

Counting on to find the difference:

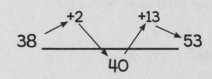

Count on from 38 to 40 and then to 53.

          2 + 13 = 15

So the difference between 38 and 53 is 15.

A Answer these.

1. $\begin{array}{r} 6\ 2 \\ -2\ 7 \\ \hline \end{array}$ 2. $\begin{array}{r} 4\ 5 \\ -1\ 8 \\ \hline \end{array}$

3. $\begin{array}{r} 5\ 5 \\ -2\ 9 \\ \hline \end{array}$ 4. $\begin{array}{r} 9\ 6 \\ -3\ 8 \\ \hline \end{array}$

B Count on to find the difference between these numbers.

1. 45 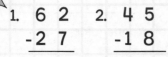 60   → difference = ____
      50

2. 19 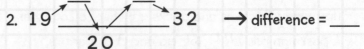 32   → difference = ____
      20

3. 27 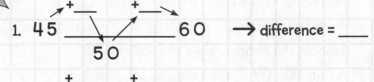 44   → difference = ____
      30

4. 86 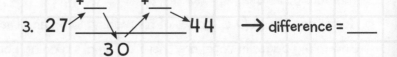 95   → difference = ____
      90

C

Read and answer these.

Use counting on to subtract mentally.

1. What is the difference between 28 and 48? _____
2. Subtract 16 from 43. _____
3. What number is 34 less than 52? _____
4. What is 80 take away 29? _____
5. How much greater is 91 than 76? _____
6. What is 66 subtract 37? _____

D

Now try this cool number puzzle!

a) 2		b)	c)	
2	d)			e)
f)		g)	h)	
	i)			j)

The puzzle works like a crossword. Solve the clues and write one digit in each space. Question (a) has been done for you.

Clues

Across	Down
a) 53 − 26	a) 44 − 22
c) 60 − 21	b) 31 − 16
d) 52 − 17	c) 59 − 23
f) 65 − 24	d) 42 − 11
h) 41 − 26	e) 60 − 15
i) 58 − 11	f) 72 − 26
j) 57 − 49	g) 64 − 27
	h) 36 − 18

Symmetry

A line of symmetry is like a mirror line.
One half of the shape looks like the reflection of the other half.

Look at these lines of symmetry.

A   Draw a line down the centre of each picture then fill in the table.

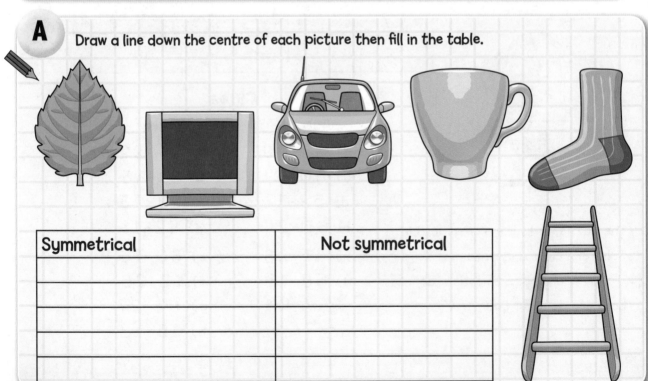

Symmetrical	Not symmetrical

symmetry When two parts of a shape are exactly the same, as though one was the reflection of the other in a mirror.

B

Complete these drawings to make symmetrical shapes.

1.
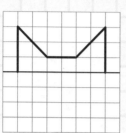

2.
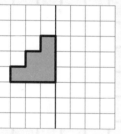

3.
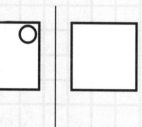

Now try to work out what are the mystery symmetry words below! Make up more mystery words of your own.

DECK CODE HOOD
V Ь Λ 1 , 1 C V , Ь I Ν

C

Some letters of the alphabet and some numbers are symmetrical. Complete each of these letters and numbers.

1. N   2. D   3. V   4. J   5. ℃

Measuring capacity

Learning objective: To read scales and use litres and millilitres.

Metric units of capacity are litres (l) and millilitres (ml).

There are 1000ml in 1l.

1000 millilitres = 1 litre

A

Write the amount shown in each jug. Look carefully at the units of measurement for each jug.

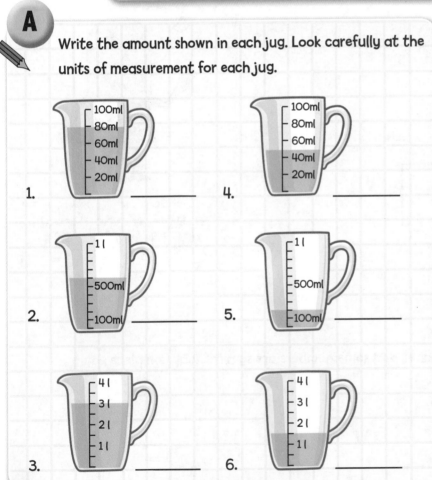

1. _____

4. _____

2. _____

5. _____

3. _____

6. _____

Amy filled a 17-litre tank with two different jugs. One jug held 3 litres and the other 4 litres. She used exactly 5 jugfuls to fill the tank. How many of each jug did she fill?

B

Each of these containers holds a different amount.

$\frac{1}{4}$ litre $\frac{1}{2}$ litre   1 litre   2 litre

How many of each of these containers would you need to fill a 1 litre jug?

1. __ cups = 1 litre   2. __ glasses = 1 litre   3. __ milk bottle = 1 litre

How many of each of these containers would you need to make 2 litres?

4. __ milk bottles = 2 litres   5. __ water bottle = 2 litres

C

Answer these.

1. How many 500ml bottles will fill a 1 litre jug? _____
2. How many 100ml large spoons will fill a I litre jug? _____
3. How many 250ml cups will fill a 1 litre jug? _____
4. How many 5ml teaspoons will fill a 100ml large spoon? _____
5. How many 500ml bottles will fill a 2 litre jug? _____
6. How many 100ml large spoons will fill a 500ml water bottle? _____

Multiplication

Learning objective: To multiply a 2-digit number by a 1-digit number.

There are different methods for multiplying numbers.

Example 1

What is 38 multiplied by 5?

$$38 \times 5 \rightarrow$$
$$30 \times 5 = 150$$
$$8 \times 5 = \underline{\ 40} +$$
$$38 \times 5 = \underline{190}$$

Example 2

What is 24 multiplied by 6?

x	20	4
6	120	24

$\rightarrow 120 + 24 = \textbf{144}$

A Complete these multiplications.

1. **14 x 4 →** 10 x 4 =
              4 x 4 = ____ +
          14 x 4 = _____

2. **25 x 9 →** 20 x 9 =
            5 x 9 = ____ +
         25 x 9 = _____

3. **37 x 6 →** 30 x 6 =
           7 x 6 = ____ +
        37 x 6 = _____

4. **58 x 3 →** 50 x 3 =
           8 x 3 = ____ +
        58 x 3 = _____

B Complete these multiplications using a grid.

1. **76 x 2 = _____**

x	70	6
2		

→ _____

2. **23 x 8 = _____**

x	20	3
8		

→ _____

3. **39 x 4 = _____**

x	30	9
4		

→ _____

4. **37 x 5 = _____**

x	30	7
5		

→ _____

multiplication This is like repeated addition.

C

Answer these. Choose a method for working out each answer. Use paper for your working out.

1. 86 x 2 = _____

2. 47 x 3 = _____

3. 19 x 9 = _____

4. 23 x 8 = _____

5. 34 x 6 = _____

6. 28 x 5 = _____

D

Read and answer these problems.

1. A bus holds 48 passengers. How many people will 4 buses hold?

_____

2. Mr Duke travels 19 kilometres each day to and from work. He works 5 days a week. How far does he travel altogether in a week?

_____

3. A market stall has 6 crates of melons. There are 35 melons in a crate. How many melons are there in total?

_____

4. A farmer fills 4 trays of eggs. Each tray holds 36 eggs. How many eggs does the farmer have?

_____

5. The battery in a mobile phone lasts 7 days. How many hours does the battery last?

_____

6. A dog eats 59 dog biscuits per day. How many will it eat in 3 days?

_____

Look for the multiplication sum in each of these problems.

Use the multiplication facts that you know.

Measuring perimeter

Learning objective: To measure the perimeter of rectangles.

The perimeter of a shape is the distance all around the edge.

Use a ruler to check these measurements.

This tile has a perimeter of 3cm + 3cm + 5cm + 5cm = 16cm

A Calculate the distance round each of these shapes. Write the perimeters in metres.

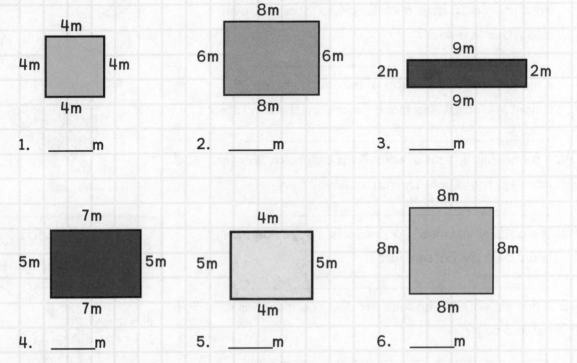

1. _____m

2. _____m

3. _____m

4. _____m

5. _____m

6. _____m

B Use a ruler to measure the sides of each rectangle.
Write the length, height and perimeter for these in centimetres.

1.

Length = _____cm

Height = _____cm

Perimeter = _____cm

2.

Length = _____cm

Height = _____cm

Perimeter = _____cm

3.

Length = _____cm

Height = _____cm

Perimeter = _____cm

4.

Length = _____cm

Height = _____cm

Perimeter = _____cm

5.

Length = _____cm

Height = _____cm

Perimeter = _____cm

6.

Length = _____cm

Height = _____cm

Perimeter = _____cm

C Complete this chart. Write the length and height of each rectangle and
calculate the perimeter.

1. 6m, 2m
2. 8m, 4m
3. 5m, 5m
4. 7m, 9m
5. 11m, 1m

Rectangle	length	add	height	Multiply total by 2	Perimeter
1	m	+	m	=	m → x 2	m
2	m	+	m	=	m → x 2	m
3	m	+	m	=	m → x 2	m
4	m	+	m	=	m → x 2	m
5	m	+	m	=	m → x 2	m

Angles

Learning objective: To use degrees to measure angles.

We use degrees (°) to measure angles.

A $\frac{1}{4}$ turn is also called a right angle.
There are 90 degrees (90°) in a right angle.

A complete turn is the same as four right angles, or 360°.

A straight line is 180°.

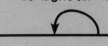

A

1. Take a piece of scrap paper.
2. Fold it to make a straight line.
3. Fold it again to make a right angle.

B Estimate the size in degrees of each of these angles.
Use your folded right angle to help. (You can fold 90° in half again to make 45°.)

1.   2.   3.   4.

5.   6.   7.   8.

Complete the table to show your estimates.

Angle	1.	2.	3.	4.	5.	6.	7.	8.
Estimated size (°)								

angle The amount by which something turns is an angle.

C

Draw the right angles on these shapes.
The first one has been done for you.

1.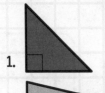

2.

3.

4.

5.

6.

D

Look at these 6 angles. Estimate the size of each angle.

Now write them in order of size, starting with the smallest: ___ ___ ___ ___ ___ ___

1. _____

2. _____

3. _____

4. _____

5. _____

6.  _____

How many right angles can you see in this shape?

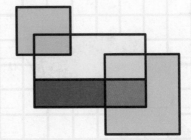

55

Division

Learning objective: To use written methods to divide.

If you know your multiplication facts it can help you to divide numbers.

Look at the trio 6, 3 and 18:

$6 \times 3 = 18$ $3 \times 6 = 18$

$18 \div 3 = 6$ $18 \div 6 = 3$

If a number cannot be divided exactly it leaves a remainder.

Example

What is 35 divided by 4?

Work out how many groups of 4 are in 35 and what is left over:

```
        8 r 3  ←————————— Answer
    4 ) 3 5
      - 3 2   (4 x 8)
      ————
        3              35 ÷ 4 = 8 remainder 3
```

Division is the opposite of multiplication.

A

Copy and complete these and find the remainders.

1.
```
      ____r__
  5 / 4 8
    __ (5 x 9)
    __
```

2.
```
      ____r__
  6 / 3 7
    __ (6 x 6)
    __
```

3.
```
      ____r__
  9 / 6 5
    __ (9 x 7)
    __
```

4.
```
      ____r__
  3 / 2 6
    __ (_ x _)
    __
```

5.
```
      ____r__
  7 / 4 0
    __ (_ x _)
    __
```

6.
```
      ____r__
  8 / 5 2
    __ (_ x _)
    __
```

B

1. ___ ÷ 4 = 7

2. 18 ÷ ___ = 2

3. ___ x 6 = 36

4. 40 ÷ 5 = ___

5. 8 x ___ = 24

6. ___ x 7 = 21

7. 54 ÷ ___ = 9

8. 6 x ___ = 48

9. 63 ÷ 9 = ___

10. ___ ÷ 4 = 8

I'm thinking of a number. It is less than 100 and if I divide it by 2, 3, 4, 5, 6 or 10 it leaves a remainder of 1. What is my number?

C

Mrs Folkes is grouping her class into teams. She has 35 pupils in her class. Read and answer these questions:

1. The School Maths Quiz has 3 pupils in each team. How many quiz teams can be made from Mrs Folkes' class? _____

2. Helper Teams have 8 children to help around the school. Any children left over will join children from another class. How many children will be left over in Mrs Folkes' class?

_____

3. Mrs Folkes is dividing her class into sports teams. Complete this chart.

Sport	Number of players in each team	Total number of teams	Number of students left over
Doubles Tennis	2 players per team	17 teams	1 left over
400m Relay Race	4 players per team	8 teams	
Basketball	5 players per team		0 left over
Volleyball	6 players per team		
Netball	7 players per team		

Fractions of quantities

Learning objective: To find fractions of numbers and quantities.

Remember that fractions have a numerator and a denominator.

$$\frac{3}{4} \leftarrow \text{numerator}$$

denominator $\rightarrow$

Example 1

What is $\frac{1}{5}$ of 40?

When the numerator is 1, just divide by the denominator.

$\frac{1}{5}$ of 40 = 40 ÷ 5 = 8

Example 2

What is $\frac{3}{5}$ of 40?

When the numerator is more than 1, divide by the denominator then multiply by the numerator.

$\frac{1}{5}$ of 40 = 8

$\frac{3}{5} = \frac{1}{5} \times 3$ ($\frac{1}{5} + \frac{1}{5} + \frac{1}{5}$)

so, $\frac{3}{5}$ of 40 = 8 x 3 = 24

A

Use the dots to work out these fractions.

1. What is $\frac{1}{3}$ of 15? _____   2. What is $\frac{1}{4}$ of 12? _____

3. What is $\frac{1}{2}$ of 18? _____   4. What is $\frac{1}{5}$ of 15? _____   5. What is $\frac{1}{3}$ of 21? _____

DEFINITION

fraction This is a part of a whole.

B

There are 24 balloons of different shapes and colours in a pack. How many of each type of balloon are there?

$\frac{1}{2}$ are red: _____ red balloons

$\frac{1}{6}$ are yellow: _____ yellow balloons

$\frac{1}{3}$ are blue: _____ blue balloons

$\frac{1}{4}$ are large balloons: _____ large balloons

$\frac{1}{8}$ are long balloons: _____ long balloons

24 ASSORTED BALLOONS

David has 64 sweets. He gives $\frac{3}{4}$ to his classmates. How many does he have left?

C

Answer each pair of questions.

1. $\frac{1}{5}$ of 35 = _____
   $\frac{2}{5}$ of 35 = _____

2. $\frac{1}{4}$ of 40 = _____
   $\frac{3}{4}$ of 40 = _____

3. $\frac{1}{7}$ of 21 = _____
   $\frac{5}{7}$ of 21 = _____

4. $\frac{1}{9}$ of 18 = _____
   $\frac{8}{9}$ of 18 = _____

5. $\frac{1}{10}$ of 70 = _____
   $\frac{7}{10}$ of 70 = _____

6. $\frac{1}{8}$ of 32 = _____
   $\frac{7}{8}$ of 32 = _____

7. $\frac{1}{3}$ of 33 = _____
   $\frac{2}{3}$ of 33 = _____

8. $\frac{1}{6}$ of 30 = _____
   $\frac{5}{6}$ of 30 = _____

Use multiplication and division facts to solve these.

59

Measuring weight

Learning objective: To read scales and use kilograms and grams.

We use **kilograms** to measure the weight of heavy objects.
We use **grams** to measure the weight of light objects.

I weigh 80 kilograms.

This salt weighs 500 grams.

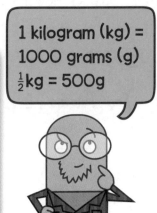

1 kilogram (kg) = 1000 grams (g)
$\frac{1}{2}$ kg = 500g

A Write the weight for each of these.

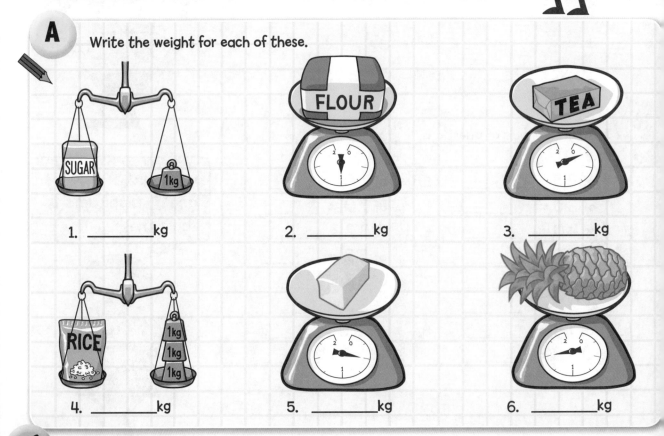

1. _____ kg

2. _____ kg

3. _____ kg

4. _____ kg

5. _____ kg

6. _____ kg

convert Change from one type of measurement to another.

B

Convert these measures.

1. 2000g = _____ kg

2. 5kg = _____ g

3. 4000g = _____ kg

4. 6kg = _____ g

5. 9kg = _____ g

6. 3000g = _____ kg

These shapes weigh 18kg altogether. If each pyramid weighs 3kg, what is the weight of each cube?

C

Write the weight of each bag to the nearest $\frac{1}{2}$kg.

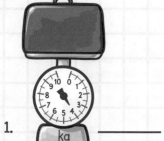

1. kg _____

2. kg _____

3. kg _____

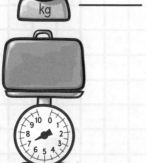

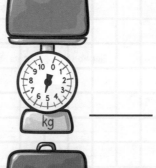

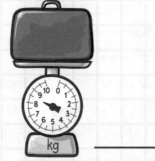

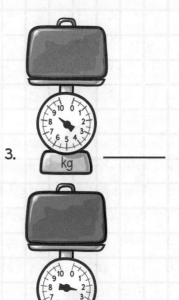

4. kg _____

5. kg _____

6. kg _____

61

Handling data

Learning objective: To read the data in bar graphs.

Do you know the names of any other types of graph?

Data is information that has been collected.

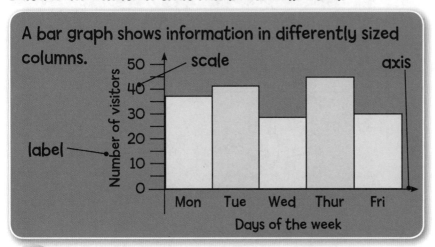

A bar graph shows information in differently sized columns.

label

A

Answer these questions about the bar graph.

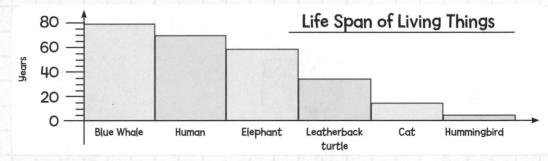

Life Span of Living Things

1. What has a life span of 70 years? _____
2. What has the longest life span? _____
3. What has the shortest life span? _____
4. How long might a leatherback turtle live? _____
5. What has a life span of about 15 years? _____
6. How much longer is a blue whale expected to live than a human? _____
7. What is expected to live twenty years more than a cat? _____
8. What creatures are expected to live longer than 50 years? _____
   _____

B

This chart shows the favourite fruits of children in Class 3.

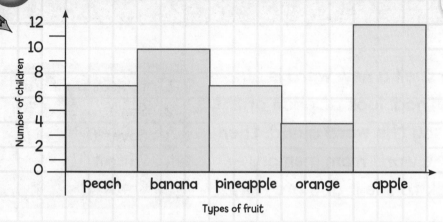

1. How many children chose bananas as their favourite fruit?
   _____ children

2. Which type of fruit did 12 children choose as their favourite?
   _____

3. Which fruit did the least number of children choose? _____

4. How many children chose peaches as their favourite fruit?
   _____ children

5. How many children altogether chose bananas or pineapples?
   _____ children

6. How many more children liked apples than bananas?
   _____ more children

7. Which two fruits did the same number of children choose?
   _____ and _____

8. Which are the two most popular fruits?
   _____ and _____

9. How many more children liked bananas than oranges?
   _____ more children

Take the difference between the number of children who liked oranges compared to those who liked peaches. Add that to the number who liked pineapple. That will give the number of people who liked my favourite fruit!

Look, say, cover, write, check

Learning objective: To learn different spelling methods.

One way to learn to spell a new word is by using this 5-step method: look at each of the letters in the word, say the word aloud, then cover it and write the word from memory – finally, you can check to see if you are right!

1. Look at the word
2. Say it
3. Cover it
4. Write it
5. Check it

You can learn to spell words in groups by looking for common letter patterns. Learn to spell these groups of words.

could, should, would

clown, frown, town

bridge, fudge, hedge

found, ground, loud, shout

coin, noise, soil, voice

cuddle, middle, little, table

A

Sometimes you can find a root word or a word within a word. Underline the root word in each of these groups of words.

<u>cook</u>	<u>cook</u>er	<u>cook</u>ery
spark	sparkle	sparkler
clear	cleared	clearly
bedroom	bedstead	bedtime
sign	signal	signature

B

Here are some tricky spellings called homophones. They are words that sound the same but are spelled differently. Write these homophones in the correct spaces below.

hear or here   ate or eight

right or write   beech or beach

would or wood   where, were or wear

What is the difference between a dog's tail and a fairytale?

1. Teri is nearly _____ years old.
2. I couldn't _____ what she said.
3. I don't know if it's the _____ way.
4. _____ you like to sleep over at my house?
5. _____ can I get the bus into town?
6. We made sandcastles on the _____ .

List any other homophones that you know:

Learn to spell tricky words by making up a mnemonic – a picture or a clue to help you remember, e.g. this hear has an ear in it!

Important spellings

You should learn to spell words that you often use when you are reading and writing. For example, days of the week, months of the year and words for numbers.

A

Write the names of the days of the week:

Today is _____ .
Yesterday was _____ .
Tomorrow is _____ .

Remember: days of the week and months of the year have capital letters.

Look for common letter strings, e.g. ember to help you spell September, November, December.

Write the names of the months of the year:

The month after April is _____ .
The shortest month is _____ .
My birthday is in _____ .

Days of the week

Monday
Tuesday
Wednesday
Thursday
Friday
Saturday
Sunday

Months of the year

January
February
March
April
May
June
July
August
September
October
November
December

Here are the words for the ordinal numbers:

first, second, third, fourth, fifth, sixth, seventh, eighth, ninth, tenth

B

Write the ordinal numbers here and try to learn them.

1st _____

2nd _____

3rd _____

4th _____

5th _____

6th _____

7th _____

8th _____

9th _____

10th _____

Learn to spell words that often appear in addresses. For example:

Street	Avenue	Close
Road	Lane	Drive

Write your home address:

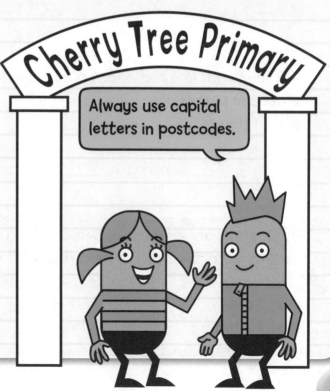

Cherry Tree Primary

Always use capital letters in postcodes.

Now write your school address:

Nouns and plurals

Learning objective: To learn plural forms of words.

A noun is a person, a place or a thing. Nouns can be singular (only one) or plural (more than one).

A

When we make most words plural we add an s to the end. Write the plurals.

sausage**s**   cake_   drink_

book_   horse_   tree_

But if a word ends in ch, sh, s, ss or x we usually add es. Write the plurals.

dish**es**   kiss__   fox__   lunch__

bus__   wish__   cross__

If a noun ends in a consonant plus y, we drop the y and write ies. Write the plurals.

pony > **ponies**   baby > _____   story > _____

daisy > _____   cherry > _____   berry > _____

These words are very tricky because they don't follow the rules. You will need to learn these plural words by heart.

man > men	child > children	leaf > leaves
mouse > mice	goose > geese	person > people

B

Read the poem below and turn the singular nouns into plural nouns. Mostly you can just add s but sometimes you have to rewrite the word in the space.

I Love the Seasons

I love it in the spring when the **bud__** burst into **million__** of tiny **flower__** .

I love it in the summer when we can make **sandcastle__** on the beach.

I love it in the autumn when the **leaf** _____ on the **tree__** turn from green to gold.

Best of all, I love it in the winter when we can make **snowman** _____ .

The words sheep, deer and fish stay the same whether they are singular or plural.

C

Change the nouns in bold in these sentences to plurals.

1. There is a **mouse** in the house!

There are _____ in the house!

2. There was only one **loaf**.

There were only two _____ .

DEFINITION

consonants The remaining 21 letters are consonants: b, c, d, f, g, h, j, k, l, m, n, p, q, r, s, t, v, w, x, y, z.

3. We saw a **goose** in the park.

We saw _____ in the park.

Prefixes and suffixes

Learning objective: To learn common prefixes and suffixes.

Prefixes are extra letters added to the beginning of words. They can change the meaning of the root word.

> For example:
>
> The rabbit appeared then disappeared then reappeared!

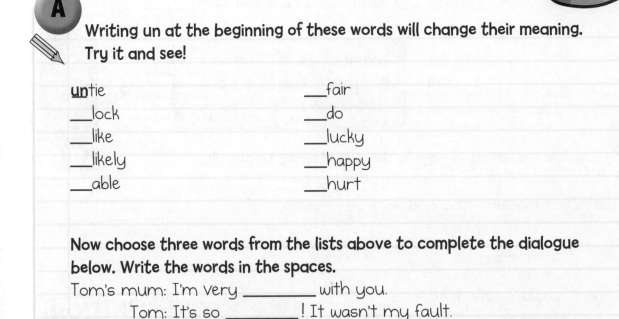

A Writing un at the beginning of these words will change their meaning. Try it and see!

<u>un</u>tie   ___fair
___lock   ___do
___like   ___lucky
___likely   ___happy
___able   ___hurt

Now choose three words from the lists above to complete the dialogue below. Write the words in the spaces.

Tom's mum: I'm very _____ with you.

Tom: It's so _____! It wasn't my fault. I'm just _____!

DEFINITION

prefixes These are the extra letters added to the beginning of a word.

B

Underline words with prefixes in this passage.
Look for dis, re, im, un.

DEFINITION

suffixes These are the extra letters added to the end of a word.

Tom and Jez went to see a remake of *Monsters of the Deep*. Writing about it in a movie review for their school magazine, they said, "The monsters were unrealistic and unimaginative really. There was lots of action but the plot was disjointed and impossible to follow."

Suffixes are extra letters added to the end of words. Look at how these suffixes change the meaning of the root word.

The word unsuccessful has a prefix *and* a suffix!

For example:

hope > hopeless > hopeful   care > careful > careless

use > useful > useless   thought > thoughtful > thoughtless

C

Choose a suffix to complete the words in these sentences.
Write less or ful.

1. It was very thought____ of Jenny to buy flowers.
2. The toy was use____ without a battery.
3. I felt hope____ at the start but then everything went wrong!
4. I knew I had to be care____ this time.

Punctuation

Learning objective: To learn basic punctuation.

Commas are used in lists to separate words and ideas.

For example:

The huge plate was piled high with bacon, egg, mushrooms, fried onions, black pudding, baked beans and tomato!

How to use commas:
- Write a comma after each item in a list.
- Write a comma to separate ideas within a sentence.

A Write the commas in these sentences.

1. We'll have two cornets with raspberry sauce a vanilla ice cream a carton of orange juice and a cup of tea please.

2. I'd like to order the tomato soup an egg and cress sandwich a banana smoothie and a chocolate muffin please.

B Write commas in these long sentences to separate the different ideas and make the text easier to read. The commas go where you pause when you read aloud.

1. The cat ran up the stairs down the corridor through the classroom and into Mrs Worgan's office!

2. Go right at the lights turn right again at the T-junction then first left.

3. The golf ball went straight down the course over the rough across the pond and landed square on the green.

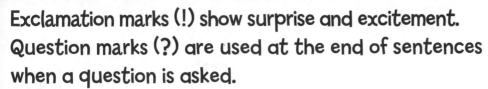

Sentences that ask questions usually begin with What, When, Where, How or Can.

Exclamation marks (!) show surprise and excitement. Question marks (?) are used at the end of sentences when a question is asked.

C

Read the sentences below and decide whether to write an exclamation mark or a question mark in each one.

1. Suddenly, all the lights went out__
2. "Aaaaaaaargh__" he cried.
3. Gina called out, "Hey, Tom__"
4. "What are 'gators__" she asked.
5. How do we know there's no life on Mars__

Speech marks highlight words that are spoken.

For example:

"How many children are coming?" asked Jason.

How to use speech marks:
- Open the speech marks at the start and close them at the end of the words spoken.
- All other punctuation goes inside the speech marks.

D

Write the speech marks in the sentences below.

1. Tara cried, Wait for me!
2. Do you think he's an elf? said Taylor.
3. Okay, said Sharon. What's wrong?
4. Wow! said Zac. You're a genius!

Apostrophes

Apostrophes can shorten words or tell you to whom something belongs.

Apostrophes are tricky! Keep practising until you understand how they work.

An apostrophe can replace missing letters:

For example:

do not > don't

it is > it's

we are > we're

they will > they'll

A

Shorten these words by using apostrophes.

cannot > **can't**

could not >_____

should not >_____

we will >_____

they will >_____

where is > _____

she is >_____

they are >_____ .

Apostrophes can also show possession.

For example: Ben's shoes.

B

Rewrite each of these phrases using an apostrophe.

1. The shoes belonging to Ben **Ben's shoes**_____

2. The book belonging to my friend _____

3. The lead belonging to the dog _____

4. The car belonging to Joe _____

5. The whiskers belonging to the cat _____

The possessive apostrophe can also tell you how many there are.

For example:

1. The boy's trainers were new. (one boy)
2. The boys' trainers were new. (two boys)

If the noun is singular the apostrophe goes before the s.
If the noun is plural the apostrophe goes after the s.

Remember these exceptions - the children's clothes, the men's clothes, the people's clothes.

C

Rewrite each of these phrases using a possessive apostrophe.

1. The fish belonging to the girl.

   The girl's fish

2. The book belonging to the teacher.

3. The television belonging to the family.

4. The red nose belonging to the clown.

5. The pram belonging to the babies.

6. The house belonging to the dolls.

7. The drawings belonging to the children.

8. The race belonging to the men.

Nouns, pronouns, connectives

Learning objective: To recognize nouns, pronouns, connectives.

A noun is a naming word. It can be a person, place or thing.

> **For example:**
>
> | The bee buzzed. bee is a noun. | Richard ran away. Richard is a noun. | The cats miaowed loudly. cats is a noun. |

A

Underline the nouns in these sentences:

The flowers were pretty.   I live in London.   The food was delicious.
Zak was asleep.   The girls laughed.   My sister has a laptop.

A pronoun is a word you can use to replace a noun so that you don't have to repeat it.

> **For example:**   Connor is kind. > **He** is kind.   **He** is a pronoun.

B

Rewrite these sentences using pronouns.
Choose from this list: him she it they them we

1. The flowers were pretty so I put the flowers in a vase.
The flowers were pretty so I put <u>them</u> in a vase.

2. Zak was asleep so I didn't want to wake Zak up.

3. I like London because London has an interesting history.

4. The girls laughed because the girls thought it was funny.

5. Chris and I went swimming. Chris and I had a great time.

Connectives are words that link ideas, sentences and paragraphs. Here are some useful connectives:

first, next, finally, consequently, later, suddenly, except, meanwhile, however, when, but, before, after, although, also, then

C Choose connectives from the list above to complete this school diary.

Taylor's school diary: Tuesday

First, after register we had a spelling test. _____ we wrote animal poems. _____ lunch, we had a visitor. It was Mrs White. She'd brought her new baby to show us. _____ lunch, we had games outside on the field. _____ , _____ it started to rain and we had to run inside. _____ , it was our science lesson. _____ , just before home time we had a story.

D

Now write your diary for yesterday in the space below. Choose connectives to link your ideas and sentences together.

Yesterday I woke up at...

77

Adjectives

Learning objective: To learn to use adjectives.

Adjectives are used to describe people, places or things.

For example:

a large dog

a small dog

A Sort these adjectives into three groups. Write them below each group heading.

aqua	average	excitable
violet	sullen	bored
indigo	raucous	scarlet
huge	angry	ginormous
miniscule	lemon	narrow

Colours: Sizes: Moods:

aqua

B Write a similar adjective (a synonym) for these common adjectives.

1. We had a nice time.
   We had a great time.

2. The pizza was okay.

3. The giant stomped his big foot.

4. It was a funny movie.

This activity is the opposite of difficult. It's easy!

DEFINITION

synonym A word with a similar meaning.
antonym A word with an opposite meaning.

Opposite adjectives are known as antonyms.

C Write an antonym for each of these adjectives.

black > white
bold > _____
hazy > _____
hairy > _____
unusual > _____

scorching > _____
expensive > _____
popular > _____
delicious > _____
polite > _____

D Change these adjectives to alter the meaning of the sentences.

1. A friendly, little dog came bounding up to her.
   A _____, _____ dog came bounding up to her.
2. It was an antique table.
   It was a _____ table.
3. It was a difficult job.
   It was a _____ job.
4. He was in a happy mood.
   He was in a _____ mood.
5. She went red when she saw him.
   She went _____ when she saw him.

Verbs and adverbs

Learning objective: To learn to use verbs and adverbs.

A verb is an action word. A sentence should have a verb.

> For example: The alien **jumped** up behind them and **burped**!

A Underline the verbs in these sentences.

1. The mouse found the cheese.

2. The cat chased the mouse.

3. The frog leaped into the pond.

4. The boy ate the chocolate bar.

5. The card skidded round the bend and crashed.

> Which sentence has two verbs?

B Change the verbs in these sentences to alter the meaning.

1. The girl dropped the ice cream.
   The girl _____ the ice cream.
2. The red team won the race!
   The red team _____ the race!
3. The family loved camping.
   The family _____ camping.
4. The children baked a cake.
   The children _____ a cake.
5. The boy ran across the road.
   The boy _____ across the road.

Remember: a verb is an action word and an adjective is a describing word.

DEFINITION

adverb A word that describes a verb or an adjective. Many adverbs end in ly.

An adverb describes the verb.

For example:

The alien suddenly jumped up behind them and burped loudly!

C Underline the verbs in these sentences then circle the adverbs.

1. The cat purred softly.

2. The giant sneezed loudly.

3. The man drove quickly.

4. The sun beat fiercely.

5. She sang beautifully.

D Change the adverbs in these sentences to alter the meaning.

1. The teacher spoke sternly.
   The teacher spoke_____ .

2. The boy carefully wrote his name.
   The boy _____ wrote his name.

3. The car quickly came to a halt.
   The car _____ came to a halt.

4. The children played noisily.
   The children played _____ .

5. I sneezed uncontrollably.
   I sneezed _____ .

DEFINITION

verb A doing or action word.
sentence A group of words that belong together.

Tenses

Learning objective: To understand different tenses.

The tense of the verbs in a sentence tells you when something happens.

It rained last night.

It's snowing now!

It will be cloudy tomorrow.

The weather forecast

Sat: sun

Sun: rain

Mon: snow

Tue: cloud

Wed: rain

A

1. What will the weather be like on Wednesday?

2. What was the weather like on Saturday?

3. What day is it today?

B

Write the past, present or future tense sentences to complete the chart.

Past	Present	Future
It was hot.	It is hot.	_____
I was hot.	_____	_____
_____	He is hot.	He will be hot.
_____	We are hot.	_____
	_____	They will be hot.

To find out more about suffixes turn to page 71.

A suffix can change the time from the present to the past:

Present	Present continuous	Past
I play.	I am playing.	I played.
I work.	I am working.	I worked.

But look what happens here. When a verb ends in a vowel and a single consonant, we double the consonant before adding the ending!

Present	Present continuous	Past
I clap.	I am clapping.	I clapped.
I stop.	I am stopping.	I stopped.

C Complete these present and past tense verbs.

Present	Present continuous	Past
I paint.	I am paint____.	I paint____.
I jump.	I am jump____.	I jump____.
I shop.	I am shop____.	I shop____.
I skip.	I am skip____.	I skip____.

These irregular verbs don't follow the usual rules. You will need to learn them by heart.

Present	Past
I drive	I drove
I hear	I heard
I sing	I sang
I get	I got
I have	I had
I go	I went

Similes and alliteration

Learning objective: To recognize similes and alliteration.

When we say something is like something else, we are using a simile.

For example: The rain is like a giant's tears.

This poem uses a list of similes to describe a cloud:

A cloud is like:
a smudge of white paint,
a fluffy pillow on my bed in the sky,
a white marshmallow,
a blob of sweet, white icing.

Try to picture the sun as something else. The sun might remind you of an orange satsuma, for example. Write each idea on a new line.

A

Try to write a list poem, using similes, based on your own thoughts about the sun.

The sun is like:

DEFINITION

simile Saying something is like something else.

Alliteration is when we put words together that start with the same sound.

DEFINITION

alliteration Words that begin with the same sounds.

For example: This monster movie is a massive hit.

B

Complete the magazine headlines below using alliteration.
Choose words from this list.

DOGS   LONG   TWOSOME   LOCKS   TERRIBLE   RECYCLE   DRAMA

REUSE AND _____
DANCING _____ **IN SCHOOL** _____
TWINS ARE A _____ _____
LOOK AFTER YOUR _____ _____

C

Complete these sentences using fun alliterations.

My alligator   is called Albert   and he's adorable.
My bear   is called Baloo   and he's big.
My c_____   is called C_____   and he's c_____ .
My d_____   is called D_____   and she's d_____ .
My e_____   is called E_____   and she's e_____ .
My f_____   is called F_____   and she's f_____ .

I'm jumping for joy!
Is that an alliteration?

85

Fiction and non-fiction

Learning objective: To distinguish between fiction and non-fiction.

Fiction books contain made-up stories. Non-fiction books contain information and fact. Fiction and non-fiction books are written in different ways.

Fiction books can have:
- dialogue
- characters
- a story or plot
- illustrations

Non-fiction books can have:
- information and facts
- photographs
- diagrams or maps
- an index

My book is called *Morris and the Aliens*. Do you think it is fiction or non-fiction?

A

Label these book titles as either F for fiction or NF for non-fiction. Write in the box next to each one.

Volcanoes ☐

Primary Science ☐

Bedtime Stories ☐

Poetry Collection ☐

The Vikings ☐

Treasure Island **F**

DEFINITION

index An alphabetical list of things in a book, with the page numbers on which each one appears, to make it easy to find things. Look at the index on page 190 of this book.

Sort your books at home into fiction and non-fiction collections.

B

Read the texts A, B and C extracted from different books and match them to the correct book titles below:

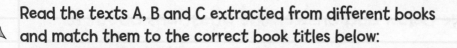

| Disappearing Worlds | Wizardy Woo | Secrets and Spies |

A. It was on the night of the next full moon that things began to go wrong. Spells that had worked perfectly well for hundreds of years had suddenly lost their magic....

From title: _____

B. Supergirl sped past the secret agents in her souped-up spy car. She had to reach Point Blank before they did. Her secret life depended on it!

From title: _____

C. The world's rainforests are vitally important to us. But every hour, thousands of square kilometres of trees are being cut down all over the world.

From title: _____

C

Which of these books would be in the fiction section and which in the non-fiction section of a library? Write the titles in the correct columns.

Fiction   Non-fiction

Fiction comprehension

Learning objective: To understand fiction text.

Read the passage below and answer the questions about it.

Sale starts today

Anxious faces peer in through the shop window.

   Inside, the manager's face is showing the wrinkles of someone twice her age as she frowns while fixing the last 'SALE!' sign on the rack. It keeps falling off and her fumbling fingers hurriedly tape it back in place.

   As the masses gather outside, like hyenas to a carcass, the shop assistants stare meekly out and dread the opening to come.

   The manager straightens out the creases from her suit and gulps before buttoning her jacket and walking towards the doors. With each step she feels like an underwater swimmer moving against the current.

   Click! The key opens the latch and she is pushed back like a leaf carried in a storm as a wave of people stampede into the shop. The doors slam open, the chaos begins.

Use the text opposite to answer the questions.

1. What do you think the story is about?

2. Who is waiting outside the shop?

3. How do you think the manager is feeling?

4. What simile is used to describe the masses gathered outside?

5. Which two similes are used to describe the manager?

Remember: a simile is when we say that something is like something else.

6. What chaos is about to begin?

7. Write an alternative title for the story.

Fables

A fable is a short story with a moral lesson. The characters in fables are often animals.

A In Aesop's fable of 'The Dog and His Bone', as retold below, some words have been left out. Predict what the words might be and write them in the spaces.

A dog was hurrying home with a big bone _____ the butcher had given him. He growled at everyone _____ passed, worried that they might try to steal it _____ him. He planned to bury the bone in the _____ and eat it later.

As he crossed a bridge _____ a stream, the dog happened to look down into _____ water. There he saw another dog with a much _____ bone. He didn't realize he was looking at his _____ reflection! He growled at the other dog and it _____ back.

The greedy dog wanted that bone, too, and _____ snapped at the dog in the water. But then _____ own big bone fell into the stream with a _____ , and quickly sank out of sight. Then he realized _____ foolish he had been.

Who was Aesop?

Aesop was a famous Greek writer of fables, who lived over two thousand years ago.

Research some other Aesop's fables at your local library.

B

Use the text opposite to answer the questions.

1. Why was the dog hurrying home?

2. Why did the other dog growl back?

3. What lesson do you think the dog learned?

4. What is the moral of the fable? Tick the correct answer, a, b or c.
   a) Waste not want not.
   b) It is foolish to be greedy.
   c) Be happy with how you look.

5. If you rewrote the fable using the same moral but a different animal character, which animal would you choose? Say why.

Classic poetry

Learning objective: To understand different types of poems.

Read this extract from 'The Pied Piper of Hamelin' by Robert Browning.

Rats!
They fought the dogs, and killed the cats,
And bit the babies in the cradles,
And ate the cheeses out of the vats,
And licked the soup from the cook's own ladles,
Split open the kegs of salted sprats,
Made nests inside men's Sunday hats,
And even spoiled the women's chats,
By drowning their speaking
With shrieking and squeaking
In fifty different sharps and flats.

A

Now answer the questions.

1. What is the extract about?

2. Look at the first three lines. Which words are alliterations - that is, begin with the same sounds?

3. Find five words in the poem that rhyme with **cats**.

4. What is a **ladle**?

5. How many cooks are there? What does the apostrophe in **cook's** tell us?

6. Why do you think the poet chose these three words: **speaking**, **shrieking** and **squeaking**?

7. What are **Sunday hats**?

8. What does the poet mean by **sharps** and **flats**?

9. If you've heard the story of the Pied Piper of Hamelin, write down what you know about it. If you're not familiar with the story, try to find a library copy.

Robert Browning was a famous writer who lived 1812—1889.

Nonsense poems and limericks

This nonsense poem tells a cautionary tale. Read it and then answer the questions below.

The Vulture
The Vulture eats between his meals
And that's the reason why
He very, very rarely feels
As well as you and I.

His eye is dull, his head is bald,
His neck is growing thinner.
Oh! what a lesson for us all
To only eat at dinner!

Hilaire Belloc

A

1. What sound is repeated three times in the first line?

2. Why did the poet choose the word **thinner**?

3. What lesson is the poet telling us?

4. Do you think this is a serious poem? Explain your answer.

Read the limerick below and answer the questions.

There was a young lady of Twickenham
Whose boots were too tight to walk quickenham.
She bore them awhile,
But at last, at a stile,
She pulled them both off and was sickenham.

Anon

Read the limerick out loud.
Which lines rhyme with which?

B

1. Why has the poet made up the words **quickenham** and **sickenham**?

2. What does **she bore them awhile** mean?

3. Write down a limerick that you know or make up one of your own.

Playscripts

Read the playscript below.

Scene 1: A New Puppy
Two dogs talking in the park.
Characters:
  Buster: Bulldog
  Sindy: Yorkshire Terrier

BUSTER: (wailing) A new puppy! After everything I've done for them.

SINDY: I knew you'd be upset. I said to our Mindy when I heard.

BUSTER: I take them for lovely walks, I eat up all their leftovers – even that takeaway muck they always dish out on a Friday… and this is the thanks I get!

SINDY: (sympathetically) You can choose your friends but you can't choose your owners.

BUSTER: What can they want a puppy for anyway?

SINDY: Well, puppies are cute.

BUSTER: Cute! Aren't I cute enough for them?

SINDY: Er…

BUSTER: Well, I'm telling you now. It's not getting its paws on my toys. I've buried them all!

How to write a playscript:
• Write the speaker's name first.
• Write each speaker on a new line.
• Describe how things are said, e.g. using adverbs in brackets.

How to write a prose story:
• Start a new paragraph for each speaker.
• Speech marks go around the words spoken.
• Use different words for 'said' e.g. replied, cried, shouted, asked, shrieked.

DEFINITION

playscript The text of a play. It includes all the words the actors would say on stage, plus a list of the cast and stage directions that help the actors to decide how to behave and move on stage.

A

Now rewrite the playscript as a prose story. Fill in the missing words.

Chapter 1: A New Puppy

"_____!" Buster wailed. "_____
_____."

"I knew you'd be upset," replied Sindy. "_____."

"I take them for lovely walks, I eat up all their leftovers - even that takeaway muck they always dish out on a Friday _____
_____!" said Buster.

"You can choose your friends but you can't choose your owners,"
_____ .

"_____?" cried Buster.

"Well, puppies are cute," said Sindy.

"_____?" replied Buster.

"Er..." said Sindy.

"Well, I'm telling you now," said Buster. "_____
_____!"

Formal letters

Read these two formal letters and answer the questions.

6 Acorn Avenue,
Newbridge,
N16 5BH.

Monday, 6 May 2009

Dear Miss Grinstead

I would be grateful if you would allow Becky to leave school early tomorrow afternoon. She has an appointment at the dentist for 3.15 pm but I would need to pick her up from school at 2.45 pm. I'm sorry that she will miss the last lesson of the day but this was the only time available.

As Tuesday is homework night, perhaps I could take Becky's homework with me when I come to collect her.

Yours sincerely

Mrs Alice Kenwood

A

1. If Becky's appointment is at 3.15 pm why does she need to leave at 2.45 pm?

2. On what day of the week is Becky's appointment?

3. Why does Mrs Kenwood apologise for taking Becky early?

4. Becky thinks she won't have to do her homework. Is this true?

DEFINITION

formal letter A business letter to someone who is not a personal friend.

Mrs A Kenwood,
6 Acorn Avenue,
Newbridge,
N16 5BH.

Monday, 6 May 2009

Botchit Kitchens,
Dead End Lane,
Newbridge,
NO1 1N.

Dear Sir

I am writing to complain about your company's shoddy workmanship on my recently fitted new kitchen.

Firstly, all of the doors are hanging off their hinges. Secondly, the drawers have been fitted upside-down so we can't put anything in them. Thirdly, you forgot to make room for the sink! What use is a kitchen without a sink?

I want to know when you are able to put these things right. Please call me to arrange a time as soon as possible.

Yours faithfully

Mrs A Kenwood

B

1. Whose address is printed on the left-hand side of the page?

2. What does 'shoddy' mean in the first sentence?

3. What is the purpose of the letter?

4. From reading the letter, how do you think Mrs Kenwood is feeling?

Instructions

Learning objective: To understand instruction text.

Read the instructions for making a thirst-quenching drink and then answer the questions below.

Apple and Raspberry Refresher

Ingredients (serves 1):
4 ice cubes
1 tablespoon raspberry syrup
250 ml or 8 fl oz apple juice
thin slices of apple for decoration

What you do:
1. Put the ice cubes in a plastic bag and crush them with a rolling pin.
2. Tip the ice into a glass.
3. Pour on the raspberry syrup.
4. Fill the glass to the top with apple juice.
5. Decorate with thin slices of apple.

A

1. What other things will you need in addition to the list of ingredients?

2. Is there enough for two glasses?

3. Is it essential that you have apple slices?

4. The instructions include the following words: put, tip, pour, fill, decorate. Are these words nouns, verbs or adjectives?

5. Write an alternative name for this drink.

Look at the recipe and read the instructions
for making a chicken salad sandwich.
Can you spot a missing ingredient?

Chicken Salad Supreme Sandwich

Ingredients:
bread
margarine
cooked chicken
lettuce leaves
tomatoes

What you do:
1. Butter the bread.
2. Put the chicken on the bread.
3. Spread on some mayonnaise.
4. Then add tomatoes and lettuce.
5. Then sandwich together.

This recipe is badly written because:
• The list of ingredients is incomplete and unhelpful,
  e.g. we don't know how much we need of anything.
• The instructions are not clear.

B

Rewrite the recipe in your own words.
Try to make big improvements on the original.

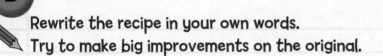

Information text

Read the text carefully and then answer the questions about it on the opposite page.

The Spanish Armada

In 1587, when Elizabeth I was Queen of England and Phillip II was King of Spain, tension between the two countries was at its greatest. Elizabeth had just signed the death warrant for the Catholic Mary Queen of Scots and this was the final straw for Phillip, who was also a Catholic.

In 1588, Phillip sent 130 warships to invade England. But the English saw the Spanish Armada arriving and the faster, smaller and more agile English ships harassed the Spanish along the English Channel. The Spanish ships were in a crescent formation curving around the English ships, and the English knew they would have to break this formation to defeat the Armada.

So the English sent burning ships to sail into the Spanish fleet. The plan worked and the Armada scattered. The Spanish ships were large, heavy and slow to manoeuvre. The English ships were quick and had better cannons so they were able to inflict a lot of damage. The Armada tried to escape back to Spain by sailing north but bad weather blew the ships on to the coasts of Ireland and Scotland.

Only half of the ships that set out in the Armada made it back to Spain. None of the English ships were lost. Elizabeth saw this victory as one of her greatest achievements.

Information text is found in non-fiction books.

A

1. Why do you think Phillip was angry when Mary Queen of Scots was executed?

2. What does 'this was the final straw for Phillip' mean?

3. Why would a crescent shape of Spanish ships be a problem for the English?

4. How did the English plan to break up the Armada?

5. What advantages did the English ships have?

6. Where was the Armada shipwrecked?

7. Approximately how many Spanish ships survived the battle?

8. How many English ships survived?

Shape and acrostic poems

Learning objective: To write a shape and an acrostic poem.

Read the shape poem.

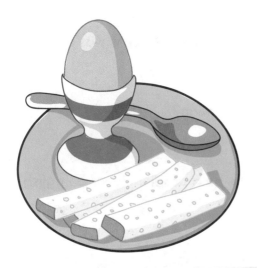

Egg

Yellow yolk
for my breakfast,
with dip-in soldiers.
I love eggy bread,
boiled, fried, scrambled,
or poached eggs... How
do you like your eggs?
"Made from chocolate,
of
course!"

How to write a **shape poem**:
- Draw an outline of a familiar object.
- Write your poem inside the outline, following the shape.
- Don't worry about rhyme - it doesn't have to rhyme.
- Try to include alliteration, e.g. yellow yolk.

This circle shape could represent a ball, a bubble, the Sun or the Moon - you decide. Then write a shape poem of your own inside the circle.

This is an acrostic poem.
The first letter in each line spells a name.

My brother
Always kicking a ball or
Running recklessly
CRASH! Into me!
OUCH! Look where you're going!

Here's another example:

    P L A Y F U L
   C **U** T E
Y E L **P**
    P A W
F U R R **y**

How to write an
acrostic poem:

- Write about something or someone that you know well.
- Spell out the subject of your poem vertically down the page.
- Alongside each letter continue with a descriptive phrase or word.

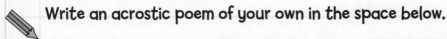

Write an acrostic poem of your own in the space below.

Settings

Learning objective: To write a description about a known setting.

A setting is the place where the events in a story happen.

A

The story below uses a girl's bedroom as a setting to begin the story. Read the opening text and answer the questions about it.

Everything matched: walls, bed, cushions, carpet, dolls – everything was either purple or pink.

Her princess bed was fluffed up with pretty pink pillows and purple sequins sparkled all around.

Rose-scented perfume filled the air and a bubbling, purple lava lamp gave off a soft, warm, purplish glow.

But the day a friend gave her a strange-looking orange ring her cosy, pink world would change forever!

1. What sort of a person would have a bedroom like this?

2. Circle any alliterations that you spot in the second paragraph.

3. What do you think she thought when her friend gave her an orange ring?

4. Do you think the ring is going to be important in the story? Say why.

Say this tongue twister:
Princess was pretty in pink!

Writers often set their stories in places that are known to them. They might be places they've visited on holiday, where they've gone to school or worked. But sometimes writers use historical settings, especially if they have an interest in history.

B

Think about a place that is well known to you. For example, it could be your bedroom, classroom or friend's room.

Write words that describe what you...
see:
hear:
feel:
smell:
taste:

Now use your notes to write a short description.

How to write a description of a place:
• Try to imagine you are there.
• Try to concentrate on the picture in your mind.
• Write down your ideas in a clear order.
• Use powerful adjectives.

Characters

Learning objective: To write a character description.

Read the character descriptions and answer the questions below.

1. Grandpa Bob, old and gnarled, like an ancient oak, sits rooted in his armchair, surrounded by his books. Age has not dulled his sense of humour or his mind, which is still as sharp and clear as ever.

2. Auntie Deera was round and plump with a soft, sunny face. When she laughed, which was often, her tummy laughed too. Her favourite saying was, "You'll never guess what happened to me today…"

3. Zak was a terrible two-year-old and a tearaway at ten. Every day at primary school, his cheeky grin got him into and out of mischief. "It wasn't me!" he'd say.

4. Charlie's blue eyes are outlined with thick, black mascara. A skull tattoo on her arm makes her look hard but I know she's not.

A

1. Which of the characters is more likely to read books: Auntie Deera or Grandpa Bob?

2. Which of the characters is most likely to enjoy food? Say why.

3. Rewrite the description of Auntie Deera using opposite adjectives to change her character.
   Auntie Deera was _____ and _____ with a _____ , _____ face.

4. Which of the characters is the youngest? Say why.

5. How old do you think Charlie is?

Try writing a description of me! What adjectives would you use?

B

Write a character description for each of these people.
Use powerful adjectives to describe their personalities and what they look like.

Miss Hanson, teacher

Bulky Bazza, weightlifter

Taz Tucker, best friend

A.T., alien being

Write a character description of someone that you know well. If you want to you can disguise their identity by changing their name, as many writers do!

How to write a character description:
• Choose names carefully because they suggest a character, e.g. Mrs Jolly.
• Ask yourself questions, e.g. What's their personality? What do they like to do?
• Different characters should speak differently, e.g. they might have favourite sayings.
• Give your character an unusual feature, e.g. eyes of different colours.

Story plans and plots

Learning objective: To learn how to plan a story.

Now it's time to write your own story. Can't think of anything to write? Don't worry, here are some ideas to try:

1. Think about all the books you have read. Choose your favourite and imitate it to help you write your own story, by changing the setting, the characters and the events.

> For example: you could write a story based on The Three Billy Goats Gruff but change the troll to a bully and the Billy Goats Gruff to you and your friends!
>
> I was walking home from school with Gaz and Tim when we saw him, swinging on the gate.

2. Retell something that has happened to you but change the characters and/or the setting.

> For example: write a story based on a time when you lost something. Perhaps you lost something very valuable belonging to someone else!
>
> Where could it have gone? I'm in big trouble now. Mum doesn't even know I had it!

3. Mix up themes from different stories. Common story themes are: good versus evil, friendship, kindness, something lost, a long journey, rags to riches.

> For example: write a story that explores two themes - friendship and rags to riches.
>
> Cindy carried her empty suitcase to the station. She pretended it was heavy so that the others wouldn't know she had nothing to put in it.

> Whatever happens in your story, the characters should be changed by it in some way. For example, an evil character might see the error of his or her ways and become a good person.

Don't try to write your story without first making a plan. Your plan might be a spider diagram, a storyboard or a written list. Look at these plans for a retelling of the Three Billy Goats Gruff.

The Troll and the Three Billy Goats (retold)

Storyboard

Billy goats teasing troll	Troll lonely	Small billy goat falls
Troll saves him	Making friends	Troll is happy

List

1. Billy Goats tease Troll.
2. Troll is lonely.
3. Smallest Billy Goat falls off the bridge.
4. Troll saves him.
5. Billy Goats make friends with Troll.
6. Now Troll is happy.

Spider diagram

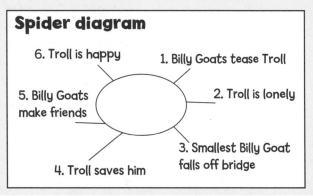

6. Troll is happy
1. Billy Goats tease Troll
5. Billy Goats make friends
2. Troll is lonely
4. Troll saves him
3. Smallest Billy Goat falls off bridge

Now try planning and writing your own story on a separate piece of paper.

Biography

Learning objective: To understand how to write a biography.

A book or a piece of writing that is an account of a person's life is called a biography.

A

The paragraphs below are all from the biography of Roald Dahl. But they are all mixed up. Read them carefully and then write the order you think they go in.

> **Biography of Roald Dahl (1916–1990)**

> 1. After school, he worked for the Shell Petroleum Company in Tanzania and in 1939, at the start of the Second World War, he joined the Royal Air Force.

> 2. He recovered and resumed duties in 1941 but then he started to suffer from headaches and blackouts.

> 3. Sadly, when he was just four, his seven-year-old sister died from appendicitis and a month later his father died from pneumonia.

> 4. He began writing in 1942 after being sent home from the army. His most popular children's books include *Charlie and the Chocolate Factory*, *James and the Giant Peach* and *The BFG*.

> 5. In 1940, Dahl was out on a mission when he was forced to make an emergency landing. Unluckily he hit a boulder and his plane crashed, fracturing his skull and his nose and temporarily blinding him.

> 6. Dahl married in 1953 and had five children.

> 7. Roald Dahl was born in Cardiff in 1916, the son of Norwegian parents.

I think that the paragraphs should go in the following order:

DEFINITION

timeline A line representing a period of time on which dates and events are marked.

B

Try and write a biography of a friend or relative. Before you start, use this space to draw out a timeline of their life. Include key events and the dates on which they happened. Write events above the line and dates below it.

Born

Now write the biography in this space:

Persuasive writing

Learning objective: To learn to write persuasively.

Read the snippets of text taken from an advertising leaflet for Awesome Towers.

AWESOME TOWERS!

- Have you got what it takes to ride the biggest rollercoaster?
- Special holiday season tickets
- Park and ride
- Gift shop
- New this year!

AWESOME TOWERS!

- A fabulous day out for all the family!
- There's something for everyone...
- Fantastic fun-packed activities for all ages.
- Ride awesome rollercoasters.
- Watch spectacular live shows.
- Enjoy an excellent choice of cafes and restaurants.
- You're guaranteed to have fun!

Next time you visit an attraction pick up a leaflet and notice the way in which it is written.

A

Now write a leaflet advertising a major attraction near you.

Amazing day out! Come rain or shine!

Holiday discount prices

Buy one ticket, get one free!

How to write persuasive text:
- Use powerful adjectives, e.g. biggest, best, terrific, exciting, guaranteed, great.
- Use active verbs, e.g. enjoy, see, find, discover, watch, ride, go, eat.
- Speak to the reader, e.g. use the pronoun 'you'.

Writing a review

Read the book review below. Notice how it is set out under different headings.

Title: Kensuke's Kingdom
Author: Michael Morpurgo

Brief outline of the story
The story is about a boy called Michael and his family who set off to sail around the world. One stormy night Michael and his dog get washed overboard. They are rescued by Kensuke, an old man who lives on a desert island.

Strengths
I liked the way the author wrote about the friendship between Michael and Kensuke. It seemed very real, like a true story.

Weaknesses
I think it was sad at the end when Michael left the island. I usually prefer stories with happier endings. But if Michael had stayed his parents would have been unhappy.

Recommendation
This is a wonderful book. Children over 8 years old would enjoy it, as I have.

Score
9/10

Write a list of your top three books, CDs and films.

DEFINITION

review An opinion or criticism of something.

A

Write a review of a book, CD or film that you have enjoyed.

Title:

Author/artist:

Brief outline of the story

Strengths

Weaknesses

Recommendation

Score

/10

How to write a review:
• Give details about the book/CD/film.
• Write about the things you liked.
• Write about the things you didn't like.
• Give a recommendation: say who would enjoy it.

Teeth

We have two sets of teeth. The first ones are called milk teeth and we begin to lose them when we are about seven. The second set are called permanent teeth and you can keep them all your life if you look after them. The purpose of the teeth is to break up food into small pieces so that we can swallow it.

Types of teeth

There are three types of teeth. They are incisors, canines and molars.

An incisor tooth is chisel shaped for cutting up soft foods such as fruit.

A canine tooth is pointed for tearing tougher food such as meat.

Molar teeth have lumpy tops, which grind together when you chew and mash up the food into tiny pieces.

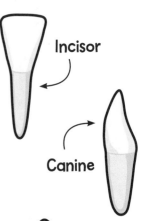
Incisor

Canine

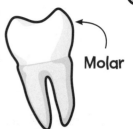
Molar

A Match the tooth with its purpose by drawing a line between them.

Tooth	Purpose
canine	cutting
molar	tearing
incisor	grinding

Look in a mirror and identify the three different types of teeth in your mouth.

The care of teeth

Learning objective: To know how to keep teeth healthy.

If you do not clean your teeth a sticky coating covers them. It is called plaque. Microbes settle in the plaque and feed on sugar in your food. They make acid. This rots the teeth. When you clean your teeth, the brush scrapes off the plaque so the microbes have nowhere to live and the toothpaste stops the work of the acid.

During the day eat hard foods such as celery, raw carrots or crunchy apples as snacks, to help keep your teeth clean.

Remember to clean the back of your teeth as well as the front. You should not eat or drink anything after you have cleaned your teeth at night because the microbes will start feeding again.

A

Choose the correct words to fill in the spaces in the sentences below.

microbes healthy clean coating teeth acid plaque

Plaque is a sticky _____ that covers the teeth. _____ live in it and make _____ that rots your _____ . When you _____ your teeth you remove the acid and _____ and keep your teeth _____ .

Food groups

Learning objective: To know that food can be sorted into groups.

There are hundreds of different foods but they can be sorted into four food groups.

Meat and fish
Lamb, beef, chicken, salmon, tuna

Fruit and vegetables
Apple, orange, pear, mango, potatoes, carrots, cabbage, onions, peas, carrots

Carbohydrates
Rice, pasta, bread

Fats and sugars
Butter, cheese, sweets, biscuits

A

To which food group does each of the foods in this meal belong? Write the name of each item and the group.

Food	Group
apple	fruit

DEFINITION

carbohydrate A substance made by plants that gives you energy. It is sometimes called starch. Rice and pasta are carbohydrates.

How the body uses food

Learning objective: To link food groups to their uses in the body.

The foods in each group help the body in a special way.

Healthy eating

To keep really healthy you need to eat different amounts of the foods in the different groups. This pyramid of food can help you remember to eat the correct amounts of food. Eat only small amounts of food at the top of the pyramid and larger amounts of the food lower down.

Fats and sugars give the body energy for action but are not good for you in large quantities.

Meat and fish help the body to grow and to repair its injuries.

Carbohydrates give the body energy for action.

Fruit and vegetables are maintenance foods. They help keep all parts of the body healthy and working well.

 A

Draw a line between each food and the way it helps the body.

Gives you energy

Keeps you healthy

Helps you grow

Do you eat healthily like the pyramid suggests?

The parts of a plant

Learning objective: To know and recognize the parts of a plant.

There are four main parts to a plant. They are the root, stem, leaf and flower.

The flower
A plant may have one or more flowers. The large brightly coloured parts of a flower are called the petals.

The leaf
A plant has many leaves. Most are green but some may have white or coloured parts.

The stem
The stems of many plants are green and bendy. The stem of a tree is made of wood and covered in bark. It is called the trunk.

The root
Roots are white and spread out through the soil.

Look at a plant in your home. Can you find all its parts?

What the plant parts do

Learning objective: To learn the purpose of each part of the plant.

Each plant part has an important task to do in the life of the plant.

The flower
The flower makes pollen which is carried away by insects or the wind. The flower also receives pollen from other flowers of the same kind and uses it to make seeds.

The leaf
The leaf makes food from the water it receives from the stem, from the air around it and from the sunlight shining on it. The food is used to make all parts of the plant grow.

The stem
The stem holds up the leaves and flowers and carries water and food to all parts of the plant.

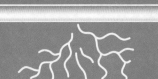

The roots
The roots hold the plant in the ground and take up water and minerals that the plant needs to make food.

 A

1. Move your finger over the picture of the plant opposite to show the way water moves through the plant.

2. Move your finger again to show how food moves.

DEFINITION

minerals Substances in the soil that the plant takes in to stay healthy.

123

Plant growth

Learning objective: To know experiments can be made to investigate plant growth.

Plant growth can be investigated by making fair tests on the effects of leaves, light, water and warmth.

How to measure growth
The length of the stem can be measured to show how a plant grows.

Experiment 1: Investigating leaves and plant growth
Two plants were given the same amount of water, light and warmth but every time one plant sprouted leaves they were carefully cut off.

Experiment 2: Investigating light and plant growth
One plant was put in a cupboard and one plant was kept on a table near a sunny window. The picture shows how the plants looked after two weeks.

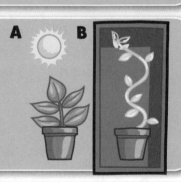

A

1. How do you think the two plants in experiment 1 compared after the experiment had run for two weeks?

2. In experiment 2, why do you think the leaves on plant B are yellow?

3. In experiment 2, why do you think plant B has grown so tall?

Have a go at the experiments on this page yourself.

DEFINITION

a fair test A test in which all the conditions except one are kept the same.

Experiment 3: Investigating water and plant growth

Five pots of cress seedlings were set up and watered each day. Different volumes of water were given to each pot as the table shows. After two weeks the plants were examined and their stems were measured.

Amount of water per day (cm³)	Stem length (cm)
5	0 (dead)
10	3.5
15	7
20	3
25	0 (dead)

Experiment 4: Investigating warmth and plant growth

A greenhouse traps the Sun's heat and makes a warm surrounding for plants. Two pots of cress seedlings were set up and the seedlings in one pot were covered with a transparent plastic cup to make a mini greenhouse. The picture shows how the pots looked after a week.

B

1. What do the results of experiment 3 show?

2. What is the effect of warmth on plant growth?

125

Introducing materials

We use many different materials to make the things we need.

Material	Use
wood	doors, furniture, bowls, spoons
brick	walls, fireplaces
stone	walls, stones
metal	pans, cutlery, cars, cans
plastic	bowls, toys, bottles, cases for computers
pottery	bowls, plant pots, ornaments
glass	windows, bottles, spectacles
cloth	clothes, blankets, towels
rubber	wellington boots, balls, tyres
paper	newspapers, books, envelopes

A

 Use the table to answer the questions.

1. Name two materials used to make walls.

2. Name three materials used to make bowls.

3. Name three materials used to make a car.

4. Which materials can you see in the picture on the right?

126

The properties of materials

Every material has some features called properties. It is the properties of a material that can make it useful to us.

Property	Examples of material with the property
hard	brick, stone, pottery
soft	cloth, plastic, foam
rough	stone, sandpaper
smooth	pottery, glass
shiny	metal, glass
dull	brick, stone
bendy	wood, rope
rigid	glass, brick

Look around you. How many of the materials in this table can you spot?

A

1. What are the properties of stone?

2. Which material is rigid, smooth and shiny?

3. Some materials do not let water pass through them. They are called waterproof materials. Other materials are not waterproof: they let water pass through them. Circle the items that are made from waterproof materials.

A   B  C  D   E   F   G

Testing wear and hardness

Learning objective: To learn how fair tests are performed on materials.

Materials can be tested for wear and hardness by performing fair tests.

Experiment 1: Test for wear

Materials can be tested for wear with a small sheet of sandpaper stuck to a piece of wood. (The wood makes the sandpaper easier to hold as it is rubbed on the materials.) The test is made fair by rubbing the sandpaper the same number of times on each material. The material is examined with a magnifying glass to observe the amount of wear.

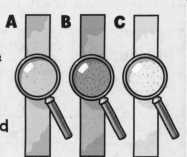

Experiment 2: Test for hardness

The hardness of a material can be tested by dropping a piece of soft modelling clay onto it. To make the test fair the ball should be dropped from the same height each time. The material which flattens a side of the ball the most is the hardest.

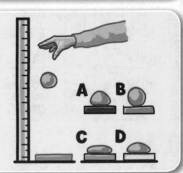

A

1. Which material in experiment 1 shows the most wear and which shows the least?

2. Write the materials A to D in experiment 2 in order of hardness, starting with the hardest.

    1. _____ (hardest)
    2. _____
    3. _____
    4. _____ (softest)

The waterproof test

Learning objective: To learn how fair tests are performed on materials.

Materials can be tested to see if they are waterproof by pouring water on them and seeing if any water passes through.

Waterproof test

Materials can be tested to see if they are waterproof by laying them on paper towels then adding the same amount of water to each one. They should all be left for the same amount of time before they are lifted off the towel. If a material is waterproof there will not be a wet mark on the towel.

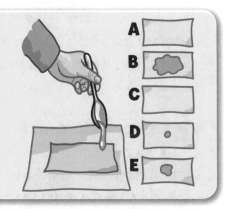

A

The paper towels A–E show the wet marks left after the waterproof test was applied to five different materials.

1. Which materials are waterproof?

2. Which materials are not waterproof?

3. Do all the materials which are not waterproof let through the same amount of water?

4. Explain your answer to question 3.

You can try all these tests yourself!

Types of rock

Learning objective: To learn about the three types of rock.

There are three types of rock. They form in different ways.

Igneous rocks
Igneous rocks form deep in the Earth where it is very hot. Basalt is black. Granite is white speckled with black and pink spots.

Sedimentary rocks
These are formed from small particles that have stuck together. Sandstone is yellow and forms from grains of sand. Limestone is grey and forms from the shells of sea creatures. Chalk is white and made from the shells of very tiny sea creatures.

Metamorphic rocks
These form from other rocks that have slipped deep into the Earth and been heated up. Slate is made from grey mud and breaks up into thin sheets. Some marble is made from crystals that shine like sugar.

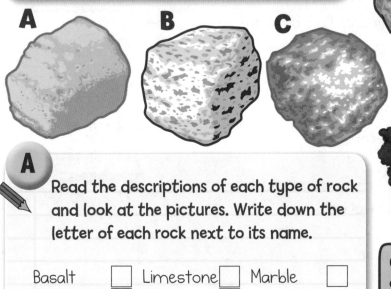

A

Read the descriptions of each type of rock and look at the pictures. Write down the letter of each rock next to its name.

Basalt ☐   Limestone ☐   Marble ☐
Granite ☐   Chalk ☐
Sandstone ☐   Slate ☐

Can you find any of these rocks around your home?

The properties of rocks

Learning objective: To learn how the properties of rocks can be tested.

Rocks can be tested for hardness and the way they react with water.

Hardness test

The hardness of two rocks can be compared by rubbing them together over white card. The rock that crumbles less is the harder.

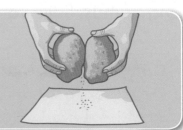

Water test

Some rocks have tiny spaces called pores inside them and let water move through them. Other rocks do not have any pores and water cannot move through them. Dry rocks can be tested by pouring a spoonful of water onto the top of them. If the rock is **porous** the water disappears into it but if the rock is **non-porous** the water remains on the top.

B

1. Look at the table and fill in the last column.

Rock	Water on rock	Porous/non-porous
Granite	Stays on top	
Limestone	Sinks into rock	
Sandstone	Sinks into rock	
Basalt	Stays on top	

2. Slate has been used to make the roof of houses. Do you think it is porous or non-porous? Why?

DEFINITION

porous Having tiny spaces in it called pores.

Soil

Soil is made from a mixture of particles of rock and humus. Soils vary from place to place.

Comparing the parts of a soil

The parts of a soil can be compared by filling a fifth of a jar with soil, nearly filling the jar with water then stirring it up for about a minute. When the water stops moving the humus floats and the rocky particles sink to the bottom. The particles at the bottom separate into layers.

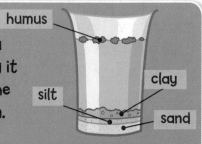

Sieving soil

A sieve can then be used to separate soil particles of different sizes. The larger particles stay in the sieve and the smaller particles pass through the holes.

For example, the sieve on the right has holes that are large enough for clay and silt to pass through but not large enough for the sand particles to pass through.

A

1. Put a ring round the rocky particles that settle down first.

   clay   sand   silt

2. Put a ring round the rocky particles that settle down last.

   clay   sand   silt

3. How do you think clay and silt could be separated?

> You could stir up potting compost and water then let it settle to compare its parts.

The drainage of soil

Learning objective: To learn how drainage is
compared in a fair test.

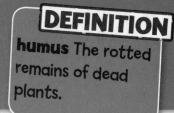

When water falls on the soil it goes down into spaces between the particles. If the spaces are large the water passes through the soil quickly and we say the soil drains well. If the spaces are small the water passes less quickly and we say the soil drains poorly.

Comparing the drainage of soils.

The drainage of different soils can be compared in the following way:

A Put the same amount of each soil in a filter funnel lined with filter paper.

B Pour the same amount of water on each soil sample.

C After five minutes measure how much water has passed through.

A

100cm^3 of water was added to four soil samples and the water was collected and measured after five minutes. This table shows the results.

Soil sample	Water added (cm^3)	Water drained (cm^3)	Water still in the soil (cm^3)
A	100	60	
B	100	43	
C	100	71	

1. How much soil was left in each soil after five minutes? Fill in the table above.
2. Which soil has the best drainage: A, B or C?

3. Which soil has the worst drainage: A, B or C?

Magnets and materials

Learning objective: To learn about magnets and magnetic
/ non-magnetic materials.

Near each end of a magnet is a place where the magnetic force is stronger. These places are called the poles of a magnet. If a magnet is placed on cork and allowed to float on water it lines up with one pole pointing to the North Pole of the Earth. This is called the magnet's north pole. The other pole of the magnet is called the south pole.

When two magnets meet.
When the poles of two magnets are brought together they may join together because they **attract** each other or they may spring apart because they **repel** each other.

Here is what happens when two magnets are brought together in three different ways:

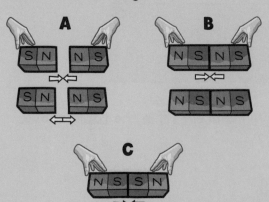

A

1. What is happening in picture A?

2. What is happening in picture B?

3. What is happening in picture C?

4. What do opposite (unlike) poles do?
   Put a ring round the answer.
   attract   repel

5. What do similar (like) poles do?
   Put a ring round the answer.
   attract   repel

Can you make fridge magnets attract and repel each other?

Testing materials

The materials in an object can be tested by bringing a magnet close to them. Only objects made of iron and steel will stick to a magnet. Iron and steel are magnetic materials. Other materials such as pottery, wood, cloth and plastic are not.

A magnet and cardboard

If a paperclip is put on one side of a piece of cardboard and a magnet on the other, the paperclip will stick to the cardboard. This happens because the force of the magnet acts through the cardboard. If the magnet is moved the paperclip on the other side moves too.

Testing the strength of magnets

The strength of a magnetic pole can be tested by attaching one paperclip after another to make a chain. The strength can also be found by placing more and more sheets of cardboard between a magnetic pole and a paperclip until the paperclip falls away.

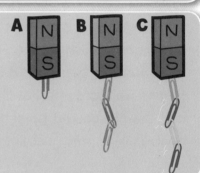

B

1. In the paperclip test above which magnet was the strongest?

2. Which was the weakest?

3. When the three magnets were tested with sheets of cardboard, which magnet do you think needed the most sheets of cardboard to make the paperclip fall away?

Springs

Learning objective: To learn that springs generate forces.

There are two types of common springs: a close-coiled spring where the coils touch each other and an open-coiled spring where the coils do not touch.

Close-coiled springs and forces

If you stretch a close-coiled spring, a tension force forms in the spring. You can feel it pulling on your fingers. This force pulls the spring back to its original length when you let go of one end. You cannot squash a close-coiled spring because the coils are already touching.

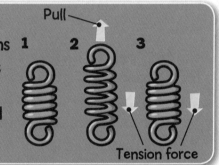

Pull

1   2   3

Tension force

Open-coiled springs and forces

When you squash an open-coiled spring a compression force forms in the spring. You can feel it pushing on your fingers. This force pushes the spring back to its original length when you let go of one end. You can also stretch an open-coiled spring just like you can a close-coiled spring.

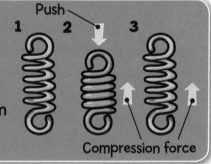

Push

1   2   3

Compression force

A

1. What will happen if a weight is put on top of spring A and then spring B?

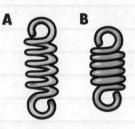

A   B

2. What will happen if a weight is hung from spring A and then spring B?

Elastic bands

Learning objective: To learn that elastic bands generate forces when stretched.

When an elastic band is stretched by a pulling force in one direction a tension force develops in the elastic band to match it. The tension force pulls in the opposite direction. If the pulling force is removed the tension force pulls the elastic band back to its original shape.

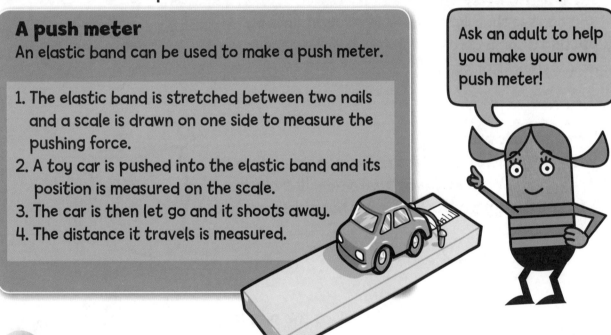

A push meter
An elastic band can be used to make a push meter.

1. The elastic band is stretched between two nails and a scale is drawn on one side to measure the pushing force.
2. A toy car is pushed into the elastic band and its position is measured on the scale.
3. The car is then let go and it shoots away.
4. The distance it travels is measured.

Ask an adult to help you make your own push meter!

A

When a toy car is pushed back along the scale to 4 it shoots away 20 centimetres. How far do you think it would go if:

1. It was pushed along the scale to 1? Circle the correct answer.

   about 1cm   about 5cm   about 30cm

2. It was pushed along the scale to 6? Circle the correct answer.

   about 1cm   about 5cm   about 30cm

Light and shadows

Light is given out by a light source. It travels in straight lines. When light reaches most objects it is stopped from travelling and a shadow forms behind the object.

How light travels

You can see that light travels in straight lines by putting a comb across a torch and shining the torch across a sheet of paper.

Why shadows form

When light strikes an object and cannot pass through it, a shadow forms on the opposite side. It is black because the light is blocked and it has a similar shape to the object.

A

Look at this picture. From which torch is the light coming to make the shadow?

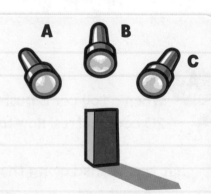

DEFINITION

light source Something that gives out light, such as the Sun, an electric lamp or a candle flame.

Materials and light

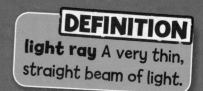

DEFINITION
light ray A very thin, straight beam of light.

Learning objective: To learn about opaque, transparent and translucent materials.

Light does not pass through most materials. However it does pass through a few.

Opaque materials stop light rays passing through them and cast dark shadows. You cannot see through them. Most materials are opaque.

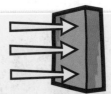

Transparent materials let most of the light rays pass straight through them. You can see clearly through them and they cast pale shadows.

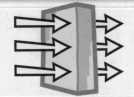

Translucent materials let some of the light rays through them but scatter them in all directions. You cannot see clearly through them and they make quite dark shadows.

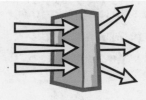

A

Put an O in the boxes of materials that are opaque, a T in materials that are transparent and a TR in materials that are translucent.

clear plastic ☐   frosted glass ☐   greaseproof paper ☐

window glass ☐   brick ☐   orange juice ☐

wood ☐   cardboard ☐

water ☐   metal ☐

The Sun and shadows

Learning objective: To know that as the Sun moves shadows change.

The Sun rises in the east, climbs in the sky until midday then slowly sinks in the west each day. As the Sun moves in the sky the shadows cast by objects change.

When the Sun is rising in the east long shadows are made which point towards the west.

When the Sun is at its highest at midday short shadows are made which point north.

When the Sun is sinking in the west, long shadows are made which point towards the east.

Remember never to look directly at the Sun. It can damage your eyes.

A

 **A** **B**  **C**

What time do you think it is at A? Circle the correct answer.

06.30   13.00   18.30

What time do you think it is at B? Circle the correct answer.

09.00   12.00   14.00

What time do you think it at C? Circle the correct answer.

06.00   10.00   17.30

How shadows change

Learning objective: To link the height of a light source with shadow length.

Scientists sometimes make models in order to experiment. In this experiment the torch is a model Sun and the block is a model of a tree.

Investigating shadow length

The torch is shone onto the block from different heights. At each height of the torch the length of the shadow is measured.

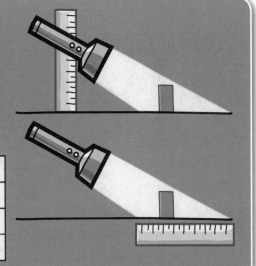

The results are recorded in a table:

Height of torch (cm)	Length of shadow (cm)
5	55
10	35
15	25

A

1. What do you think the length of the shadow might be when the torch is at a height of 25 centimetres? Circle the correct answer.

    5cm   10cm   15cm

2. Write T (true) or F (false) next to each statement.

    A   As the torch rises the shadows get longer.   ☐

    B   As the torch rises the shadows get shorter.   ☐

    C   As the torch sinks the shadows get longer.   ☐

Try this experiment with your own torch.

The human skeleton

Learning objective: To learn about the parts of the human skeleton.

The human skeleton has 206 bones. They work together to support the body and help it move. Some bones protect some of the organs of the body too.

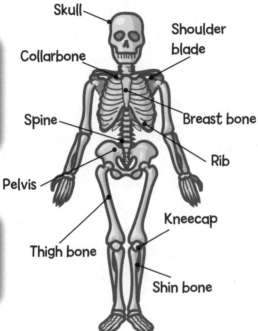

Skull

Shoulder blade

Collarbone

Spine

Pelvis

Breast bone

Rib

Kneecap

Thigh bone

Shin bone

The skull is made from a group of bones that protect the brain.

The shoulder blade and collar bone attach the arm bones to the spine.

The ribs form a cage which protects the heart and lungs and moves to help us breathe.

The pelvis attaches the leg bones to the spine.

A

Name the bones that connect the toe bones to the skull.

Can you find and feel the bones labelled on the diagram on your own body?

DEFINITION

organ A body part which performs a particular task in keeping the body alive. For example, the heart is an organ which pumps blood around the body.

Animal skeletons

Learning objective: To learn that animals have different types of skeletons.

Many animals have bony skeletons but some have skeletons made of shell, a horn-like material or even water.

Bony skeletons

Fish, amphibians, reptiles (scaly-skinned animals), birds and mammals (animals with hair) have bony skeletons with a skull and a spine. Fish have bones that support their fins, and reptiles such as snakes do not have arm or leg bones.

Animals with a skeleton of armour

Some animals have a skeleton made from hard material on the outside of their bodies. Crabs and shrimps have a skeleton made of shell. Insects and spiders have a skeleton made from a substance with properties similar to horn and finger nails.

Animals with a water skeleton

Earthworms and slugs have spaces in their bodies that are full of water. These spaces act like a water skeleton and support the body.

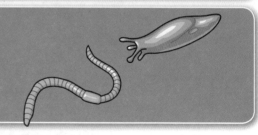

A

Match the animal to its skeleton by drawing lines between them.

slug

frog   skeleton of bone

shrimp   skeleton of armour

spider   water skeleton

snake

DEFINITION

amphibian An animal with a smooth skin, which has a tadpole stage in its life e.g. a frog.

Movement

Muscles can contract themselves, but they cannot stretch themselves. A contracting muscle can give a bone a pull and this can make part of the body move. To make the body part move back to its original position you need a second muscle.

Pairs of muscles

There are pairs of muscles all over the body that contract, move body parts and help each other stretch. One example of a pair of muscles are those in the upper arm. They are called the biceps and the triceps.

1

The biceps contracts and pulls on the bones of the lower arm to raise them.

The triceps is stretched by the contracting biceps.

2

The triceps contracts and pulls on the bones of the lower arm to lower them.

The biceps is stretched by the contracting triceps.

Stretch out your right arm. Spread out the fingers of your left hand and press them into your biceps. Bend your right arm. What do you feel with your fingers?

DEFINITION

contract To make shorter.

144

Exercise

Learning objective: To learn that exercise helps to keep the body healthy.

DEFINITION

joints The places where bones are joined together such as at the elbow or the knee.

Exercise is any activity that makes the body move about. Exercise helps the body stay healthy.

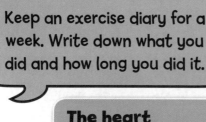

Keep an exercise diary for a week. Write down what you did and how long you did it.

Exercise keeps bones and joints strong. It makes the muscles strong, too.

The heart
The heart is a bag of muscle that pumps the blood round the body. The blood takes food and oxygen to all parts of the body. When the muscles exercise they need more food and oxygen and the heart pumps faster to provide them. This makes the heart muscles stronger, too.

Exercise and fat
The body stores energy from food as fat. Too much fat makes the body so heavy that it strains the muscles and heart to move. Exercise uses up energy and stops the body getting overweight.

A

Name five ways in which exercise makes the body healthy.

Did you do more exercise or less exercise than you thought? Do you need to do more?

145

The size of habitats

Learning objective: To know that organisms live in habitats.

The place where an organism lives is called its habitat.
There are different sizes of habitat.

Micro habitat
This is the smallest type of habitat. It could be the place under a stone where a centipede lives or the underside of a leaf, which is home to greenfly.

Mini habitat
This is made up from many micro habitats. A bush is a mini habitat. It has micro habitats, which include the bark where beetles live, leaves where caterpillars feed, flowers where spiders may hide to catch insects and roots where eelworms gather to feed.

Habitat
This is the largest habitat and is made up from micro and mini habitats. A forest is a habitat that is made up from trees, bushes, grassy areas and bare ground covered with dead leaves.

A

 A    B    C   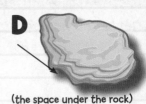 D

(the space under the rock)

Identify A–D by putting a ring around your answer.

What is A?	habitat	mini habitat	micro habitat
What is B?	habitat	mini habitat	micro habitat
What is C?	habitat	mini habitat	micro habitat
What is D?	habitat	mini habitat	micro habitat

DEFINITION

organism A living thing such as a plant or an animal.

The conditions in a habitat

Learning objective: To know that organisms in a habitat are adapted to their conditions.

There are many different kinds of habitat. Each one has a set of conditions, which organisms must be adapted to if they are to survive there.

The forest habitat
The trees make the forest shady, so only plants that are adapted to growing in dim light such as ferns and mosses can grow on the ground there.

The pond habitat
The poorly draining soil causes a deep pool to form. Only organisms adapted to living in water can survive there.

The rocky shore habitat
Only organisms which are adapted to the beating of the waves and living in pools of salty water can survive in this habitat.

The swellings called bladders on the seaweed make most of the plant float and stay in the light to make food. The seaweed has a holdfast, which looks like a root, to grip the rock.

The limpet has a sucker to grip the rock and its shell stops it losing water when the tide is out.

The crab breathes in salty water and can hide away when the tide comes in.

A

1. Meadow grass is adapted to growing in bright light. Explain what would happen to it if it was planted in a wood and why.

2. Describe what would happen to a limpet if it could not grip onto the rocks.

Grouping living things

Learning objective: To use the features of organisms to put them into groups.

Every organism has features, which can be used to put it into a group with other organisms. Organisms are placed in groups because it makes them easier for scientists to study.

Here are some major groups of animals and their features:

Animal group	Features
Insects	Six legs
Spiders	Eight legs
Fish	Scales and fins
Amphibians	Smooth slimy skin
Reptiles	Scaly skin
Birds	Feathers
Mammals	Hair or fur

You could also check out the definition of **amphibian** on page 143.

A

1. Write the letter of each animal next to the group it belongs to.

Animal group	Animal
Insects	
Spiders	
Fish	
Amphibians	
Reptiles	
Birds	
Mammals	

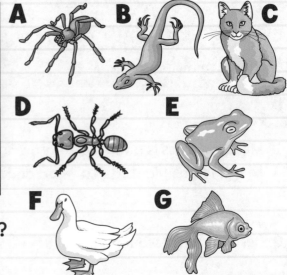

2. Which animal group do humans belong to?

Keys

Learning objective: To learn how to use a key to identify organisms.

A key is a number of features about organisms, which is set out in a series of questions. As each answer is made you move onto the next question until you find the identity of the organism.

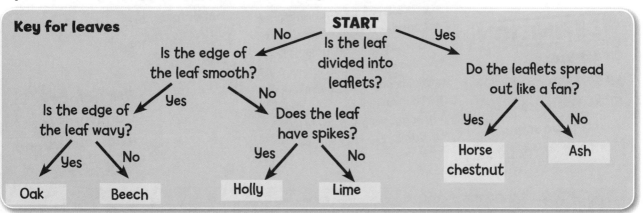

Key for leaves

START
Is the leaf divided into leaflets?

No → Is the edge of the leaf smooth?

Yes → Do the leaflets spread out like a fan?

Is the edge of the leaf smooth?
Yes → Is the edge of the leaf wavy?
No → Does the leaf have spikes?

Is the edge of the leaf wavy?
Yes → Oak
No → Beech

Does the leaf have spikes?
Yes → Holly
No → Lime

Do the leaflets spread out like a fan?
Yes → Horse chestnut
No → Ash

For example, in this key if you look at leaf **F** below and begin to answer questions about it you will see that it is not divided into leaflets, that its edge is not smooth and it does not have spikes. This means that the leaf belongs to a lime tree.

A

Use the key to identify the leaves.

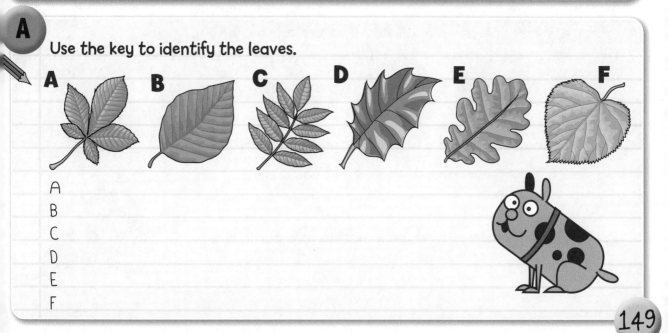

A B C D E F

A

B

C

D

E

F

Food chains

Learning objective: To understand food chains.

Some organisms in a habitat make food and others eat it.
Scientists use special words for organisms that make and eat food.

DEFINITION

producers
Organisms that make food are called producers. Plants are producers.

consumers Organisms that eat food are called consumers. All animals are consumers. There are two kinds of consumer. A prey animal is one that is consumed by another animal.
A predator is an animal that consumes another animal.

herbivores
Herbivores eat plants and are prey animals.

carnivores
Carnivores are meat-eaters and predators.

Organisms in a food chain

1. The first organism in a food chain is a plant. It makes food from air, sunlight and water.
2. The second organism is a herbivore. It is probably also a prey animal.
3. The third animal in a food chain is a predator.

Food chains can have more than three organisms in them. For example, if an eagle ate a fox it could be added to the food chain. This would mean that the fox is a prey animal as well as a predator, but it would still remain a carnivore.

A

1. Draw a food chain with a human acting as a herbivore.

2. Draw a food chain with a human acting as a carnivore.

Note that the arrow always goes from the food to the feeder.

Protecting habitats

Learning objective: To understand the significance of protecting habitats.

DEFINITION

habitat The place where an animal or plant lives.

Plants and animals come to live together in a habitat because they are adapted to the conditions there. If these conditions change the organisms are less well adapted and may die or move away.

When a forest is cut down

The trees provide mini habitats for many organisms so when the trees are removed the organisms are removed too. Predators such as woodpeckers that feed on insects in the bark do not have anything to eat and move away. The trees also provide shade so when they are removed strong light can reach the ground. Plants such as ferns are less adapted to these conditions and may die.

When a pond is drained

Animals such as fish, which cannot move away when the pond is drained, die. Frogs may hop away to find a new pond. Birds such as herons, which feed on fish and frogs, move away too. Water plants cannot survive in dry soil and die. Birds, which nested among the water plants, move away.

A

Imagine that there are plans to drain a pond and clear forest near your home. Write a letter or design a poster that protests about the plan and encourages people to conserve the habitats.

Find out why the Earth's rainforests need protecting and make a poster about it.

Temperatures

Learning objective: To learn how to use a thermometer and record temperatures.

A thermometer is used to measure temperature. The temperature is measured in degrees Celsius. Scientists write this as °C.

The parts of a thermometer

A thermometer has a glass tube with a swelling at one end called the bulb. There is a red- or green-coloured liquid, called alcohol, in the bulb and lower part of the tube. The alcohol is the same temperature as the air around it. This temperature can be found by looking at the top of the alcohol in the tube and reading the scale next to it.

This thermometer shows that the air around it is 20°C.

Taking the temperature

When taking the temperature of a liquid the thermometer bulb must be kept in it for a few moments before the reading is made. This gives the alcohol time to become as cold or as hot as the liquid. The thermometer bulb must be kept in the liquid while the scale is being read.

The thermometer shows the temperature of the water to be 25°C.

A

1. Shade in thermometer A to show a temperature of 30°C.

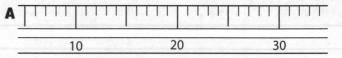

2. Shade in thermometer B to show a temperature of 25°C.

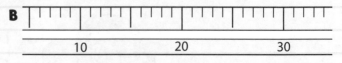

3. Shade in thermometer C to show a temperature of 33°C.

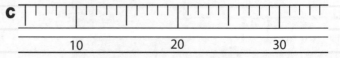

4. Shade in thermometer D to show a temperature of 39°C.

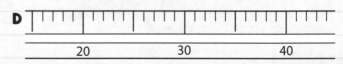

The temperatures of substances can change over a short period of time. These changes can be measured with a thermometer and a clock.

Recording temperature change

A cup of warm water was left to stand on a table for four minutes. The temperature of the water was taken five times to show how the temperature changed in this time.

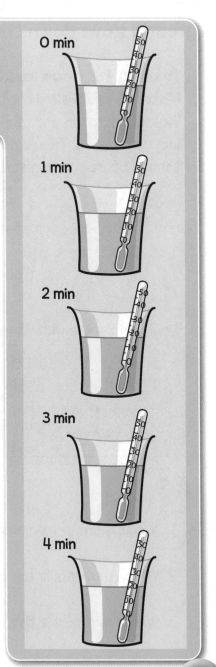

0 min

1 min

2 min

3 min

4 min

B

1. Fill in this table from the data in the pictures.

Time from start (minutes)	Temperature (°C)

2. Why do you think the last two temperatures were the same?

3. Now make a line graph from the data in the table.

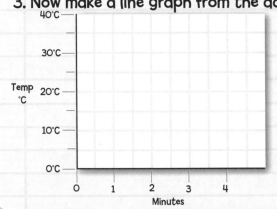

Heat insulators

Learning objective: To compare how materials prevent heat moving using a fair test.

Heat moves from a warmer place to a cooler place. A material which greatly slows down the movement of heat is called a heat insulator.

Testing for heat insulation

Materials are tested for their heat-insulating property in the following way. Plastic cups and their lids are covered with the materials but one cup and lid are left uncovered. The same amount of warm water is poured into each cup and the temperature is taken straight away and ten minutes later.

Wool   Aluminium foil   Cotton

A   B   C   D

The table shows how the temperature changed in each cup:

Time (Mins)	Cup A Temp (°C)	Cup B Temp (°C)	Cup C Temp (°C)	Cup D Temp (°C)
0	50	50	50	50
10	30	45	32	40

 A

1. What is the purpose of the cup without material?

2. Which material is the best heat insulator?

3. Which material is the worst heat insulator?

Heat conductors

DEFINITION

melting Changing from a solid into a liquid.

Learning objective: To compare how quickly materials move heat using a fair test.

A material which heat can move through quickly is called a heat conductor. When some materials, like butter, become hot they melt. This means that the melting of butter can be used as a rough measure for the movement of heat.

Testing for heat conduction

A test for heat conduction can be made by taking a plastic, wooden and steel spoon of the same size and placing them in a bowl of hot water. A small lump of butter is then added to the handle of each one and the lumps are watched for signs of melting. The lump of butter melts fastest on the spoon made out of the best heat conductor.

Look at cooking-pan handles and explain why they are not made out of metal.

A

1. Which spoon is made from the best heat conductor, A, B or C?

2. If you made a spoon out of aluminium foil and used it in this in this experiment explain what would you think would happen.

3. How does the metal in cooking pans help food in the pans to cook?

155

Solids and liquids

Learning objective: To compare solids and liquids and measure their volumes.

The shape of solids and liquids

A solid has a fixed shape. This means its shape does not change if you leave it for a long time. It does not change its shape if you turn it over or put it on its side. Liquids do not have any shape. They take up the shape of any container they are poured into.

The volume of solids

If a solid is shaped like a block you can easily find its volume by measuring its height, width and length then multiplying them.

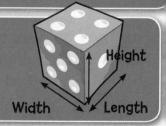

Measuring the volume of a liquid

The volume of a liquid can be found by pouring it into a measuring cylinder, which is set on a flat horizontal surface. The eye should be brought level with the liquid surface for reading the volume on the scale of the cylinder.

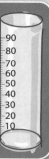

Take a block of wood and find its volume. Then look at another block of wood, estimate its volume then measure it. How good was your estimate?

A

1. What is the volume of liquid in each measuring cylinder?

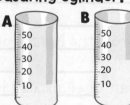

Volume in A:

Volume in B:

2. If the liquid from A is poured into B what will be the volume of liquid in B?

Flowing

Learning objective: To compare runniness in liquids and powdered solids.

Different liquids flow with different speeds. Their speeds can be compared in a runniness test. Tiny solid particles can also flow like liquids.

Comparing runniness
The runniness of liquids can be compared by letting them flow down a ramp and timing how long they take to reach the bottom.

Can you make salt or flour flow like a liquid?

Grains and powders
Some pieces of solid such as sand and salt are so small they form particles called grains. Some solid particles such as pepper and flour are even smaller and form powders. When grains or powder particles are put together into a group they slide over each other. This makes them flow and they can be poured like a liquid. They do not form a real liquid because they cannot form drops.

A

1. How is the runniness test made fair?

Pour water very slowly out of a jug and look for drops. What happens when you pour flour in the same way?

2. Here are the results when the runniness test was done on three liquids, A, B and C. Which one is water, which one is treacle and which one is vegetable oil? Fill in the table.

Liquid	Time to flow (seconds)	Identity of liquid
A	8	
B	20	
C	6	

Melting

If any solid is heated up enough it melts and becomes a liquid.

Melting

Solids keep their shapes at normal temperatures but if they are heated they may become soft and start to sag. If they are heated more strongly they flow and can form drops. When this happens they have turned into a liquid. The process in which a solid changes into a liquid is called melting. It is a reversible change.

For example, a candle is made from wax which melts when it gets hot and turns into a liquid.

A

Here are the melting points of four substances:

Substance	Melting point (°C)
A	200
B	750
C	430
D	850

1. The four substances are heated up in a furnace. What is the order in which they melt?

2. Which substances are still solid at 500°C?

3. When A melts, how much hotter must the furnace get before D melts?

DEFINITION

reversible change A change that can be reversed because it has an opposite. The opposite of melting is freezing.

Freezing

Learning objective: To know that when liquids are cooled down enough they freeze.

If any liquid is cooled down enough it freezes and becomes a solid.

Freezing

When a liquid freezes it cannot flow any more and takes on a fixed shape. It becomes a solid. Freezing is a reversible change.

For example, to make an ice lolly you put liquid water in a container in the freezer. As the water cools down it freezes. It becomes a solid and takes on the fixed shape of the container.

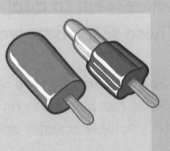

Freezing point

The temperature at which a liquid freezes is called its freezing point. This is the same as the melting point of the solid that it forms.

For example, the freezing point of the liquid called water is 0°C and is the same as the melting point of the solid, called ice, which it forms.

Different substances freeze at different temperatures. They do not all freeze at 0°C. The molten wax running down the side of a candle freezes when its temperature drops to 50°C.

What is the opposite of freezing?

A

Choose the correct words to fill in the spaces in the sentences below.

| hot | solid | side | liquid | 600°C | flowing | freezing | away |

The molten rock _____ down the _____ of a volcano is a _____ and may be as _____ as 1200°C. It cools as it moves _____ and when it reaches _____ it stops flowing and becomes _____ rock. The _____ point of the rock is 600°C.

159

Dissolving

When many solids are stirred up with water, they simply fall to the bottom of the liquid and form a layer. They do not dissolve. But some solids seem to disappear into the water when they are stirred up in it. These solids dissolve in the water.

Solids that do not dissolve
Sand, marbles, chalk and glass beads do not dissolve in water. They are said to be insoluble in water and form a layer at the bottom of the water container.

When a solid dissolves
When a solid such as sugar or salt dissolves in water it splits up into very tiny particles called atoms and molecules. They are so small that you would need a very powerful microscope to see them. The mixture of the water and the solid dissolved in it is called a solution.

Coloured solid, coloured solution
Some solids such as instant coffee granules are coloured.
When they dissolve they colour the water, too.

A

1. Which of these substances dissolve in water? Tick the boxes.
   chalk ☐   sugar ☐   instant coffee granules ☐
   salt ☐   marble ☐   sand ☐
2. Loose tea leaves turn hot water brown but they also form a layer at the bottom of the cup or teapot. Explain what is happening.

DEFINITION

atoms The small particles from which everything is made. They can bunch together to form small groups called molecules.

Filtering

Learning objective: To understand how filtering can be used to separate materials.

A filter is a material with holes in it which can let a liquid pass through.

Filtering sand and water

When you pour water mixed with sand into a filter paper two things happen. The sand is held back in the paper because the grains are too big to go through the holes but the water passes through.

When salt water is filtered

When salt water is poured into a filter the salt does not settle in the paper. It is in such small pieces in the water that they pass through the holes with the water. A dissolved substance cannot be separated from water by filtering.

A

1. Some sand and sugar have been stirred into water. Explain what will happen to the sand, the sugar and the water when they are poured into a filter paper.

2. How could you tell that salt or sugar has been stirred into a cup of water if you cannot see it?

3. Is the paper in a tea bag a filter? Explain your answer.

The force meter

Learning objective: To learn how to use a force meter.

A force meter is made out of a close-coiled spring (see page 136). It is used to measure the strength of pulling forces. A force meter may also be called a spring balance or a newton meter.

The parts of a force meter

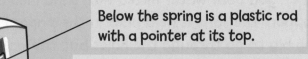

At the top of the force meter is a handle that can be attached to a hook or held in the hand.

Inside the force meter is the stretching part – a close-coiled spring.

Below the spring is a plastic rod with a pointer at its top.

Around the spring and rod is a plastic case with a scale on it. The units on the scale are newtons.

The plastic rod is attached to a hook. This can be attached to weights or items which are to be pulled.

How to measure a force

When the hook is pulled the spring stretches and the pointer moves down inside the case. When the pointer stops moving the size of the force can be measured by looking at the where the pointer has stopped on the scale.

A

What size of force does each of these force meters read?

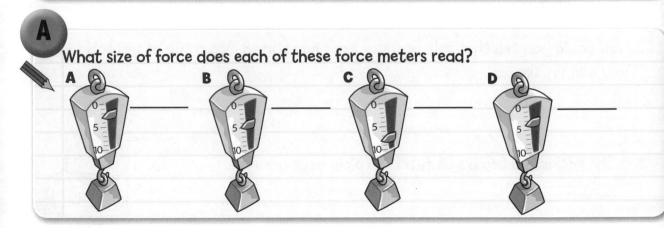

Friction

Learning objective: To learn about friction.

When you try and push your foot along a floor you can feel the force of friction.

Friction is a force generated when two surfaces touch and one is pushed or pulled over the other. Friction acts in the opposite direction to the pull or push. If the push or pull is weak the force of friction will match it in strength and the surfaces will not slide. When the push or pull reaches a certain strength it overcomes the force of friction and the surfaces move.

Push   Friction

A force meter and friction

A force meter can be used to find the force needed to overcome the force of friction.

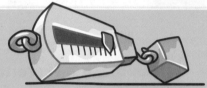

A slope and friction

When a wood block is placed on a tray and one end raised slightly, gravity pulls on the block and friction holds the block in place. At some point, as the end is raised, gravity overcomes friction and the block slides. The height of the slope at which the block starts to slide is used to measure the force of friction.

Gravity

Friction

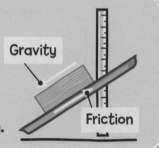

A

When the slope method was used to test the friction between an object and different materials on the slope the following readings were made:

Make a slope like the one on this page, put different shoes on it and tip it up. Which shoe has the best grip?

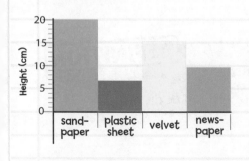

Arrange the materials in the order of the strength of friction between them and the block, strongest first.

1.

2.

3.

4.

Water resistance

Learning objective: To learn about water resistance and how to compare it.

When an object moves through water the water pushes back on it. This pushing force is called water resistance.

Streamlined shapes

A shape which allows water to flow over it easily is called a streamlined shape. Water resistance to this shape is low so the shape moves fast through the water. A spindle shape is a streamlined shape.

Comparing different shapes

The water resistance of different shapes of modelling clay can be compared using a tall cylinder of water and a stopclock.

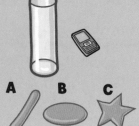

A piece of modelling clay is divided into three lumps of the same size then each lump is made into a different shape. Each shape is dropped down the cylinder and the time taken for it to fall to the bottom is recorded.

A

1. Which shape do you think went fastest and which went slowest?

2. How could the results be checked?

DEFINITION

spindle A rod which comes to a point at each end.

Air resistance

Learning objective: To learn about air resistance and how to compare it.

Air resistance is a pushing force which acts on an object as it moves through the air. The larger the surface moving through the air, the stronger the pushing force on the object.

Running holding up a small piece of card is easy...

...but running holding up a big piece of card like this is hard! Try it and see!

Comparing air resistance

Model parachutes of different sizes can be used to investigate air resistance. They are all dropped from the same height of about 1.5 metres and the time for them to reach the ground is timed.

A Three parachutes were made with square canopies and dropped from a height of 1.5 metres. Here are the results:

Parachute	Length of canopy side (cm)	Time of fall (seconds)
A	10	5
B	8	4
C	12	6

1. Which parachute took longest to fall?

2. Which parachute took the shortest time?

3. How long would you expect a parachute with 14-centimetre sides to fall? Circle the correct answer.   9 seconds   8 seconds   7 seconds

Electrical components

Learning objective: To recognize the different components in an electrical circuit.

DEFINITION

circuit A loop made by joining electrical components together so that a current of electricity can flow through them.

The items that make up an electrical circuit are called components. Here are the components that are used in simple circuits in science experiments. You must never use mains electricity for experiments.

Battery
This provides the electricity for use in circuits. It is sometimes called a cell. The button at one end is called the positive terminal and the base at the other end is called the negative terminal. The power of the battery is measured in volts (V). Batteries for science circuits should be 1.5 V.

Wires
These are made of metal and may be coated in plastic. They conduct electricity around the circuit. Some wires have crocodile clips on their ends.

Switch
This controls the flow of electricity in a circuit.

Switch open

Switch closed

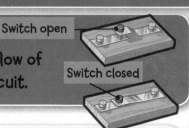

Motor
When electricity flows through the motor it makes the motor shaft spin.

Buzzer
This makes a sound when electricity flows through it.

Lamp
This has a thin wire called a filament which lights up when electricity flows through it.

A

Match the component to its use by drawing lines between them:

motor   conducts electricity

wire   controls the flow of electricity

buzzer   provides movement

switch   makes a sound

Making a circuit

Learning objective: To recognize when a circuit will conduct electricity.

All electrical circuits need one or more batteries to provide the electric current. They need wires to connect the components together and a switch to control the flow. They also need either a lamp, buzzer or motor to tell you when the current is flowing.

Making connections

A circuit is made by connecting up the components. The connections need to be made carefully because if a gap is left between components or the components can easily separate a current will not flow. A gap between components can occur at any place in a circuit.

Electricity will only flow in a circuit when all the connections are firmly made and the switch is switched on.

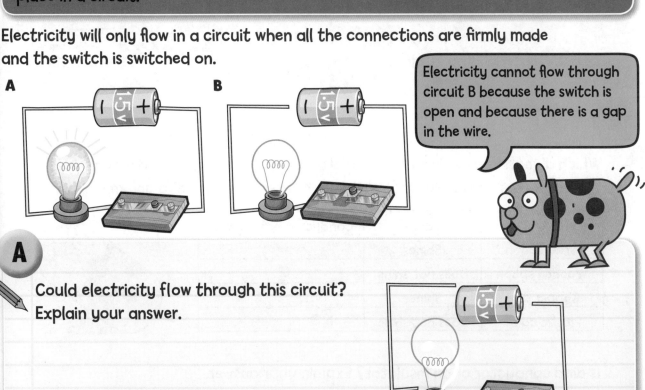

A

B

Electricity cannot flow through circuit B because the switch is open and because there is a gap in the wire.

A

Could electricity flow through this circuit? Explain your answer.

Testing for flow

Learning objective: To test materials to find out if they conduct electricity.

Materials which let electricity flow through them are called electrical conductors. Materials which do not let electricity flow through them are called electrical insulators. A simple circuit can be used to test them.

Setting up the test circuit

Materials can be tested in this circuit. A material to be tested is placed across the gap and the switch is switched on. If the lamp lights, electricity is flowing around the circuit and through the material. This means that the material is a conductor. If the lamp does not light when the switch is switched on, the material is not allowing electricity to flow. This means the material is an insulator.

Place material to be tested here.

A

1. Which objects are made from materials that are conductors and which from materials that are insulators? Write your answers in the third column.

Object	Lamp	Conductor or insulator
Steel spoon	Shines	
Wooden spoon	Does not shine	
Copper pipe	Shines	
Plastic comb	Does not shine	

2. Is air a conductor or an insulator? Explain your answer.

Switches

Learning objective: To make switches from simple materials.

A switch has two pieces of metal called contacts. When the switch is off, the metal contacts do not touch and the air between them acts as an insulator so a current of electricity does not flow. When the switch is on, the contacts touch and electricity flows round the circuit.

If you have a torch, look and see if you have to press or slide the switch to complete the circuit and turn the torch on.

The importance of switches
You do not **have** to use a switch to make a circuit. It is possible just to wrap the ends of the wires together to complete a circuit. However it is better to use a switch because it does not wear out the ends of the wires and is quicker to use.

Paperclip switch
This type of switch is turned on when the paperclip is moved so it can touch both drawing pins.

Pressure switch
This type of switch is turned on when the two inside surfaces are pressed together. The two pieces of foil touch and let electricity flow. It can be used in a burglar alarm circuit. The switch could be under a mat so that it turns on if it is stepped on.

A

Explain how the switch on the right works.

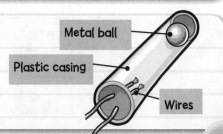

Metal ball

Plastic casing

Wires

Using more batteries

A battery pushes a current of electricity around a circuit. Its pushing power is measured in volts (see page 166). A bulb is made to take a certain voltage of electricity. If it is too high the lamp will burn out.

Lining up batteries

A battery pushes electricity around a circuit from its negative terminal to its positive terminal. If another battery is added it must be placed with its positive terminal next to the negative terminal of the first battery. If batteries are lined up with both their positive or negative terminals together no current will flow.

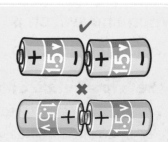

Batteries and lamps

1. When two 1.5V batteries are correctly placed in a circuit their pushing power on the electricity rises to 1.5V + 1.5V = 3.0V. This makes a lamp shine more brightly.
2. Most lamps are made to work at a voltage of 3V or less. This means that they will only work with a maximum of two 1.5V batteries. If a third 1.5V battery is added it will raise the pushing power to 4.5V and this will burn out the lamp.
3. The voltage at which a bulb will work is marked on the casing. Some larger lamps have a voltage of 4.5V and can be used with three batteries in a line.

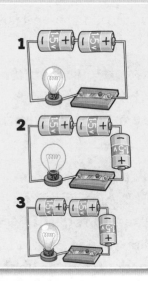

A

Which pairs of batteries allow electricity to flow? Tick the boxes.

 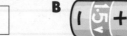

Using more lamps

Learning objective: To learn how the arrangement of lamps affects their brightness.

The wire in the filament of the lamps offers some resistance to the flow of the current. As the current pushes against this resistance it can make the filament so hot that it shines and comes 'on'. The way lamps are connected together in a circuit affects the way they shine.

Lamps in a row

When lamps are arranged in a row they are said to be in **series**. If two lamps are in series the resistance of one filament wire adds to the resistance of the second and the current is slowed down. This means two lamps do not shine as brightly as a single lamp.

Lamps side by side

Lamps can be arranged side by side by connecting each one to the battery and switch. The lamps arranged in this way are said to be in **parallel**. The resistance of one filament does not add to the resistance of the other filament in this arrangement so both lamps shine as if they were in the circuit on their own.

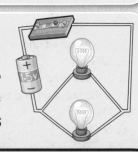

A

One of arrangements A, B and C shines more dimly than the other two. Which is it?

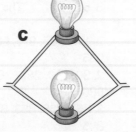

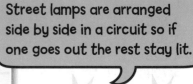

Street lamps are arranged side by side in a circuit so if one goes out the rest stay lit.

Answers

Pages 10-11

A Odd → 15, 17, 29, 33, 21
   Even → 24, 20, 32, 22, 16, 38, 30
B Check odd numbers are circled
   and even numbers are underlined.
   1. 11   2. 12   3. 17   4. 16 5. 21
C
   1. 36, 38   2. 27, 29
   3. 52, 54   4. 33, 35
Challenge: 234,456,789 is odd.

Pages 12-13

A
   1. 300 + 90 + 8
   2. 200 + 10 + 7
   3. 400 + 50 + 2
   4. 600 + 80 + 3
   5. 100 + 60 + 5
   6. 700 + 0 + 9
B
   1. nine hundred and forty-one
   2. 326
   3. five hundred and thirty-four
   4. eight hundred and seventy
   5. 219
   6. 650
C
   1. 83
   2. 545
   3. 911
   4. 628
   5. 704
   6. 367

Pages 14-15

A
   1. 147 is less than 152.
   2. 476 is less than 479.
   3. 753 is more than 735.
   4. 521 is more than 381.
   5. 190 is less than 390.
6. 214 is less than 244.
7. 586 is more than 585.
8. 592 is more than 497.
B
   1. 264 > 254
   2. 328 < 431
   3. 190 > 119
   4. 536 > 523
   5. 708 < 807
   6. 655 > 635
C
   1. 112, 125, 159, 191
   2. 278, 373, 387, 483
   3. 622, 645, 668, 739
   4. 410, 416, 460, 461
   5. 704, 743, 760, 778
   6. 195, 309, 459, 815

Pages 16-17

A
   1. 42, 47   5. 84, 80
   2. 93, 95   6. 479, 469
   3. 43, 40   7. 668, 768
   4. 60, 64   8. 531, 431

B
   1. 85, 145   The rule is +15
   2. 500, 450   The rule is -50
   3. 943, 743   The rule is -100
   4. 619, 419   The rule is -100
   5. 820, 840   The rule is +5
   6. 472, 502   The rule is +10
   7. 113, 103   The rule is -10
   8. 512, 527   The rule is +3
C
   1. 995, 985, 975, 965, 955, 945,
   935
   2. 80, 180, 280, 380, 480,
   580, 680
   3. 857, 855, 853, 851, 849, 847,
   845

Pages 18-19

A
   1. cone   4. cuboid
   2. sphere   5. cylinder
   3. cube   6. prism

B
   1. All cylinders. The odd one out
   is a cone.
   2. All cubes. The odd one out is
   a sphere.
   3. All cuboids. The odd one out
   is a prism.

C

Name of shape	cube	cuboid	prism
Total number of faces	6	6	5
Number of square and rectangle faces	6	6	3
Number of triangular faces	0	0	2

D
   1. never 2. always 3. never
   4. sometimes

Pages 20-21

A
   1. 7 + 8 = 15   15 - 7 = 8
       8 + 7 = 15   15 - 8 = 7
   2. 6 + 6 = 12   12 - 6 = 6
       6 + 6 = 12   12 - 6 = 6
   3. 9 + 5 = 14   14 - 9 = 5
       5 + 9 = 14   14 - 5 = 9
   4. 9 + 7 = 16   16 - 9 = 7
       7 + 9 = 16   16 - 7 = 9
B
   1. 15, 150, 1500
   2. 4, 40, 400
   3. 12, 120, 1200
   4. 2, 20, 200

C

1. 9
2. 8
3. 18
4. 4
5. 60
6. 900
7. 50
8. 200

D.

1. 90
2. 14
3. 7
4. 600
5. 700
6. 40

Pages 22-23

A

1. 10
2. 30
3. 30
4. 50
5. 160
6. 160
7. 180
8. 190

B

1. 300 2. 300 3. 400 4. 500

C

1. 40+30=70
2. 100-50=50
3. 80-40=40
4. 150+40=190
5. 500+200=700
6. 800-300=500
7. 700+200=900
8. 800-600=200

Pages 24-25

A

1. 67
2. 57
3. 35
4. 29
5. 98
6. 78

B

1. 76
2. 78
3. 82
4. 43
5. 81
6. 89

C

55 + 20 → 72 + 3
71 + 8 → 29 + 50
37 + 30 → 63 + 4
62 + 6 → 48 + 20
74 + 5 → 39 + 40

D

1. 37
2. 42
3. 54
4. 64
5. 46
6. 84

Pages 26-27

A

1. 42
2. 84
3. 23
4. 28
5. 63
6. 12
7. 42
8. 62

B

1. 95
2. 52
3. 73
4. 42
5. 24
6. 41

C

1.

IN	56	78	27	49	15	64
OUT	52	74	23	45	11	60

2.

IN	65	91	42	77	59	83
OUT	35	61	12	47	29	53

Pages 28-29

A

1. 6.3
2. 0.9
3. 12.4
4. 18.5
5. 11.1

1. $\frac{8}{10}$
2. $7\frac{2}{10}$
3. $16\frac{7}{10}$
4. $20\frac{6}{10}$
5. $4\frac{9}{10}$

B

1. 2
2. 20
3. $\frac{2}{10}$
4. $\frac{2}{10}$
5. $\frac{2}{10}$
6. 2

C

7.5   8.1   8.7   9.7
12.9   13.4   13.8   14.3

D

1. >
2. <
3. >
4. <
5. >
6. >
7. <
8. <

Pages 30-31

A

1. quadrilateral, 2. octagon,
3. hexagon, 4. pentagon,
5. triangle, 6. hexagon

B

1. Hexagons - odd one out is a quadrilateral.
2. Quadrilaterals - odd one out is a pentagon.
3. Ovals - odd one out is a circle.

C Check all shapes are pentominoes.

Pages 32-33

A

1. (d) $\frac{2}{4}$
2. (a) $\frac{2}{8}$
3. (b) $\frac{3}{9}$
4. (d) $\frac{2}{10}$

B

1. $\frac{5}{10} = \frac{1}{2}$
2. $\frac{1}{3} = \frac{2}{6}$
3. $\frac{4}{16} = \frac{1}{4}$

C

Possible answers could be:

$\frac{1}{3} = \frac{2}{6} \frac{3}{9} \frac{4}{12} \frac{5}{15} \frac{6}{18}$
$\frac{1}{4} = \frac{2}{8} \frac{3}{12} \frac{4}{16} \frac{5}{20} \frac{6}{24}$
$\frac{1}{5} = \frac{2}{10} \frac{3}{15} \frac{4}{20} \frac{5}{25} \frac{6}{30}$

Pages 34-35

A Estimates should be within 1cm of the answers to section B.

B

1. 5cm
2. 8cm
3. 10cm
4. 7cm
5. 9cm
6. 3cm

Answers

C

1. 9cm	5. 12.5cm
2. 4cm	6. 3.5cm
3. 2.5cm	7. 6cm
4. 10cm	8. 8cm

Pages 36-37

A

1. 24	4. 45
2. 21	5. 30
3. 27	6. 56

B

1. 35, 42	5. 9, 18
2. 40, 48	6. 14, 28
3. 60, 54	7. 16, 32
4. 80, 72	8. 18, 36

C

1. 24	4. 2
2. 27	5. 5
3. 3	6. 3, she will have some left over.

D

2 x 9 → 18   4 x 7 → 28
x   x   x   x
6 x 9 → 54   9 x 3 → 27
↓   ↓   ↓   ↓
12   81   36   21

Pages 38-39

Answers may be given as digital (12.10) or in the 'o'clock' format.

A

1. 2.10	5. 7.30
2. 4.50	6. 10.15
3. 1.20	7. 3.45
4. 6.05	8. 8.35

B

1. 6.45
2. 8 o'clock
3. 20 minutes

4. 4.25
5. 45 minutes

C

1. 9.05	4. 9.50
2. 9.15	5. 9.55
3. 9.30	6. 10.10

Pages 40-41

A

1. 9 square centimetres
2. 5 square centimetres
3. 7 square centimetres
4. 6 square centimetres
5. 10 square centimetres
6. 6 square centimetres

B

1. 8 square metres
2. 49 square metres
3. 16 square metres
4. 23 square metres
5. 25 square metres

C Check designs are all different and made from 8 squares.

Pages 42-43

A

1. 200 + 70 + 15 = 285
2. 100 + 70 + 18 = 188
3. 200 + 70 + 11 = 281

B

1. 180	2. 244
3. 296	4. 161

C

1. 93	4. 171
2. 83	5. 294
3. 72	6. 251

D

1. 90km	4. 185
2. 67	5. 160
3. 61	6. 174cm

Pages 44-45

A

1. 35	2. 27
3. 26	4. 58

B

1. 5 + 10 = 15	2. 1 + 12 = 13
3. 3 + 14 = 17	4. 4 + 5 = 9

C

1. 20	4. 51
2. 27	5. 15
3. 18	6. 29

D

a) 2	7	b) 1	c) 3	9
2	d) 3	5	6	e) 4
f) 4	1	g) 3	h) 1	5
6	i) 4	7	8	j) 8

Pages 46-47

A

Symmetrical → leaf, TV, ladder
Not symmetrical → car, cup, sock

B

Check each shape drawn is an exact reflection.

C

1. M	4. U
2. B	5. 3
3. X	

Challenge: DECK, CODE, HOOD, WHAT, TOW, HIM

Pages 48-49

Challenge: Three 3-litre jugs and two 4-litre jugs.

A

1. 80 ml	4. 50ml
2. 600ml	5. 200ml
3. 3 litres	6. 1.5 litres

B

1. 4   2. 2   3. 1   4. 2   5. 1

C

1. 2	4. 20
2. 10	5. 4
3. 4	6. 5

Pages 50–51

A

1. 56   2. 225
3. 222   4. 174

B

1. 152   2. 184
3. 156   4. 185

C

1. 172   4. 184
2. 141   5. 204
3. 171   6. 140

D

1. 192   4. 144
2. 95km   5. 168 hours
3. 210   6. 177

Pages 52–53

A

1. 16m   4. 24m
2. 28m   5. 18m
3. 22m   6. 32m

B.

1. l 3cm, h 1.5cm, p 9cm
2. l 4cm, h 1cm, p 10cm
3. l 1.5cm, h 1.5cm, p 6cm
4. l 2.5cm, h 3cm, p 11cm
5. l 4cm, h 1.5cm, p 11cm
6. l 2.5cm, h 1cm, p 7cm

C

Rectangle	length	add	height	Multiply total by 2	Perimeter
1	6m	+	2m	= 8 m → x 2	16m
2	8m	+	4m	= 12 m → x 2	24m
3	5m	+	5m	= 10 m → x 2	20m
4	7m	+	9m	= 16 m → x 2	32m
5	11m	+	1m	= 12 m → x 2	24m

Pages 54–55

A Check instructions have been followed.
B Check estimated angles are close to these:
   1. 150°   2. 110°   3. 90°   4. 45°
   5. 30°   6. 130°   7. 70°   8. 90°
C Check all right angles are marked correctly.
D The order should be:
   6 (30°), 4 (45°), 2 (60°),
   1 (100°), 3 (120°), 5 (160°)
Challenge: 32

Pages 56–57

A

1. 9 r 3   4. 8 r 2
2. 6 r 1   5. 5 r 5
3. 7 r 2   6. 6 r 4

B

1. 28   6. 3
2. 9   7. 6
3. 6   8. 8
4. 8   9. 7
5. 3   10. 32

C

1. 11 teams
2. 3 children
3.

17 teams	1 left over
8 teams	3 left over
7 teams	0 left over
5 teams	5 left over
5 teams	0 left over

Challenge: 61

Pages 58–59

A

1. 5   4. 3
2. 3   5. 7
3. 9

B 12 red, 4 yellow, 8 blue, 6 large, 3 long

C

1. 7, 14   5. 7, 49
2. 10, 30   6. 4, 28
3. 3, 15   7. 11, 22
4. 2, 16   8. 5, 25

Challenge: 16

Pages 60–61

A

1. 1kg   4. 3kg
2. 1kg   5. $\frac{1}{2}$kg
3. $\frac{1}{4}$kg   6. 1$\frac{1}{2}$kg

B

1. 2kg   4. 6000g
2. 5000g   5. 9000g
3. 4kg   6. 3kg

C

1. 4kg   4. 7kg
2. 5$\frac{1}{2}$kg   5. 8$\frac{1}{2}$kg
3. 9 kg   6. 2$\frac{1}{2}$kg

Challenge: 4kg

Pages 62–63

A

1. human 2. blue whale
3. hummingbird 4. 35 years
5. cat 6. 10 years
7. leatherback turtle
8. elephant, human and blue whale

B

1. 10 2. apple 3. orange 4. 7
5. 17 6. 2 7. peach and pineapple 8. apple and banana
9. 6

Challenge: Bananas are my favourite fruit.

Answers

Pages 64-65

A

spark, sparkle, sparkler
clear, cleared, clearly
bedroom, bedstead, bedtime
sign, signal, signature

B

1. eight 2. hear 3. right 4. Would 5. Where
6. beach

Pages 66-67

A

The month after April is May.
The shortest month is February.

B

first, second, third, fourth, fifth, sixth, seventh,
eighth, ninth, tenth

Pages 68-69

A

sausages, cakes, drinks, books, horses, trees
dishes, kisses, foxes, lunches, buses, wishes, crosses
ponies, babies, stories, daisies, cherries, berries

B

buds, millions, flowers, sandcastles, leaves, trees,
snowmen

C

1. There are mice in the house! 2. There were only
two loaves. 3. We saw geese in the park.

Pages 70-71

A

untie, unlock, unlike, unlikely, unable, unfair, undo,
unlucky, unhappy, unhurt
I'm very unhappy with you. It's so unfair! I'm just
unlucky!

B

Tom and Jez went to see a remake of Monsters of
the Deep. Writing about it in a movie review for
their school magazine, they said, "The monsters
were unrealistic and unimaginative really.
There was lots of action but the plot was

disjointed and impossible to follow."

C

1. thoughtful 2. useless 3. hopeful 4. careful

Pages 72-73

A

1. We'll have two cornets with raspberry sauce, a
vanilla ice cream, a carton of orange juice and a
cup of tea please.
2. I'd like to order the tomato soup, an egg
and cress sandwich, a banana smoothie and a
chocolate muffin please.

B

1. The cat ran up the stairs, down the corridor,
through the classroom and into Mrs Worgan's
office!
2. Go right at the lights, turn right again at the
T-junction, then first left.
3. The ball went straight down the course, over
the rough, across the pond and landed square on
the green.

C

1. Suddenly, all the lights went out!
2. "Aaaaaaaargh!" he cried.
3. Gina called out, "Hey, Tom! "
4. "What are 'gators?" she asked.
5. How do we know there's no life on Mars?

D

1. Tara cried, "Wait for me!"
2. "Do you think he's an elf?" said Taylor.
3. "Okay," said Sharon. "What's wrong?"
4. "Wow!" said Zac. "You're a genius!"

Pages 74-75

A

can't, couldn't, shouldn't, we'll, they'll, where's,
she's, they're

B

1. Ben's shoes 2. My friend's book 3. The dog's lead
4. Joe's car 5. The cat's whiskers

C1. The girl's fish 2. The teacher's book 3. The

family's television 4. The clown's red nose 5. The babies' pram
6. The dolls' house 7. The children's drawings 8. The men's race

Pages 76-77

A

The flowers were pretty.
Zak was asleep.
I live in London.
The girls laughed.
The food was delicious.
My sister has a laptop.

B

1. The flowers were pretty so I put them in a vase.
2. Zak was asleep so I didn't want to wake him up.
3. I like London because it has an interesting history.
4. The girls laughed because they thought it was funny.
5. Chris and I went swimming. We had a great time.

C

First, after register we had a spelling test. Then we wrote animal poems. Before lunch, we had a visitor. It was Mrs White. She'd brought her new baby to show us. After lunch, we had games outside on the field. But (or However) suddenly it started to rain and we had to run inside. Next, it was our science lesson. Finally, just before home time we had a story.

Pages 78-79

A

Colours: aqua, lemon, scarlet, violet, indigo. Sizes: huge, ginormous, miniscule, average, narrow. Moods: sullen, raucous, excitable, angry, bored.

B

Possible answers: 1.We had a great time. 2.The pizza was average. 3.The giant stomped his huge foot. 4.It was a hilarious movie.

C

black > white
bold > weak (or shy)

scorching > freezing
expensive > cheap

hazy > clear   popular > unpopular
hairy > bald (or hairless)
delicious > horrible (or disgusting)
unusual > common (or normal)   polite > rude

D

Possible answers:
1. A fierce, big dog came bounding up to her.
2. It was a new/modern table.
3. It was an easy job.
4. He was in a sad/bad mood.
5. She went white when she saw him.

Pages 80-81

A

1. found 2. chased 3. leaped 4. ate 5. skidded, crashed (two verbs).

B

Possible answers:
1. The girl grabbed the ice cream.
2. The red team lost the race!
3. The family hated camping.
4. The children bought a cake.
5. The boy walked across the road.

C

1. The cat purred softly
2. The giant sneezed loudly
3. The man drove quickly
4. The sun beat fiercely
5. She sang beautifully.

D

Possible answers:
1. The teacher spoke angrily.
2. The boy hurriedly wrote his name.
3. The car slowly came to a halt.
4. The children played happily.
5. I sneezed loudly.

Pages 82-83

A

1. It will rain on Wednesday. 2. It was sunny on Saturday. 3. Today it is Monday.

B

Past	Present	Future
It was hot	It is hot	It will be hot.

Answers

I was hot.   I am hot.   I will be hot.
He was hot.   He is hot.   He will be hot.
We were hot. We are hot.   We will be hot.
They were hot. They are hot. They will be hot.

C

I am painting   I painted
I am jumping   I jumped
I am shopping   I shopped
I am skipping   I skipped

Page 85

B

REUSE AND RECYCLE
DANCING DOGS IN SCHOOL DRAMA
TWINS ARE A TERRIBLE TWOSOME
LOOK AFTER YOUR LONG LOCKS

Pages 86-87

A

Volcanoes – NF
Primary Science – NF
Bedtime Stories – F
Poetry Collection – F
The Vikings – NF
Treasure Island – F

B

A. Wizardy Woo B. Secrets and Spies C.
Disappearing Worlds

C

Fiction: Wizardy Woo, Secrets and Spies
Non-fiction: Disappearing Worlds

Page 89

A 1. It is about the first day of the sales.
2. The customers are waiting outside.
3. She is nervous, dreading the opening.
4. They are 'like hyenas to a carcass'.
5. She is 'like an underwater swimmer moving against the current', and
'like a leaf carried in a storm'.
6. There will be chaos as the customers fight for bargains.
7. Any suitable title.

Pages 90-91

A

A dog was hurrying home with a big bone that the butcher had given him. He growled at everyone who passed, worried that they might try to steal it from him. He planned to bury the bone in the garden and eat it later.

As he crossed a bridge over a stream, the dog happened to look down into the water. There he saw another dog with a much bigger bone. He didn't realize he was looking at his own reflection! He growled at the other dog and it growled back. The greedy dog wanted that bone too, and he snapped at the dog in the water. But then his own big bone fell into the stream with a splash, and quickly sank out of sight. Then he realized how foolish he had been.

B 1. He was hurrying to get home quickly before someone stole the bone from him.
2. The other dog growled back because it was just a reflection.
3. The dog learned that he had lost his bone because he was greedy.
4. b) it is foolish to be greedy.

Pages 92-93

A 1. The extract tells us about the rats.
2. killed/cats/cradles; bit/babies.
3. vats, sprats, hats, chats, flats.
4. A ladle is a big spoon.
5. The apostrophe tells us that there is only one cook.
6. They all begin with an 's' sound. They also rhyme.
7. Hats worn on Sunday when going to church.
8. Sharps and flats are the sounds made by the rats squeaking and shrieking.

Pages 94-95

A 1. The 'ee' or 'ea' sound is repeated three times.
2. Thinner rhymes with dinner. When we eat we usually grow fatter so growing thinner is a sign that the Vulture isn't well.
3. The poet is telling us that we shouldn't eat between meals.
4. It is not a serious poem but a fun or nonsense poem. The Vulture doesn't have a bald head and thin neck for the reasons given in the poem.

B

1. Quickenham and sickenham are both made-up words, invented to rhyme with Twickenham.
2. She wore the boots for a short time before she took them off.

Page 97
A

"A new puppy!" Buster wailed. "After everything I've done for them."

"I knew you'd be upset," replied Sindy. "I said to our Mindy when I heard."

"I take them for lovely walks, I eat up all their leftovers – even that takeaway muck they always dish out on a Friday... and this is the thanks I get!" said Buster.

"You can choose your friends but you can't choose your owners," said Sindy, sympathetically.

"What can they want a puppy for anyway?" cried Buster.

"Well, puppies are cute," said Sindy.

"Cute! Aren't I cute enough for them?" replied Buster.

"Er ...," said Sindy.

"Well, I'm telling you now," said Buster. "It's not getting its paws on my toys. I've buried them all!"

Pages 98-99
A

1. She needs to leave at 2.45 pm in order to get to the appointment on time.
2. Becky's appointment is on Tuesday.
3. Mrs Kenwood apologises because Becky will miss lessons.
4. Becky will have to do her homework because Mrs Kenwood is going to collect it.

B

1. The kitchen company's address is on the left.
2. Shoddy means careless and of poor quality.
3. Mrs Kenwood wants the company to put right these mistakes.
4. She is exasperated, disappointed and annoyed.

Page 100
A

1. You need a plastic bag, rolling pin, glass.
2. No there isn't enough for two glasses because the ingredients state 'serves 1'.
3. No, because the apple slices are for decoration only.
4. These words are verbs.
5. Any appropriate name.

Page 103
A

1. Phillip was angry because she was a Catholic and he was too.
2. 'This was the final straw' means that Phillip could not take any more.
3. The English ships could then be attacked on three sides.
4. The plan was to sail burning ships into them.
5. The English ships were smaller, faster, more agile and had better cannons.
6. The Armada was shipwrecked off the coast of Ireland and Scotland.
7. About 65 Spanish ships survived.
8. They all survived.

Page 106
A

1. A girl, about 9 years old. Someone who loves girlie things.
2. Princess, pretty, pink, pillows and purple; sequins and sparkled.
3. She was disappointed the ring wasn't pink.
4. The ring is important because we are told that it would change her world forever.

Page 108
A

1. Grandpa Bob is more likely to read books.
2. Auntie Deera is most likely to enjoy food because we are told she is plump.
3. (possible answer) Auntie Deera was tall and thin with a sharp, sullen face.
4. Zak is the youngest because the text implies that he is ten years old.
5. Charlie is probably a teenager.

179

Answers

Page 112
A 7, 3, 1, 5, 2, 4, 6

Page 118
A

canine – tearing
molar – grinding
incisor – cutting

Page 119
A

Plaque is a sticky **coating** that covers the teeth. **Microbes** live in it and make **acid** that rots your **teeth**. When you **clean** your teeth you remove the acid and **plaque** and keep your teeth **healthy**.

Page 120
A

chicken leg – meat and fish
chips and peas – fruit and vegetables
bread and butter – fats, starches and sugar
apple – fruit and vegetables
bread – carbohydrates

Page 121
A

fish – helps you grow
bread – gives you energy
orange – keeps you healthy
If you eat more fatty and sugary food than meat , fruit and vegetables you are not eating healthily.

Page 123
A

1. You should move your finger from the root upwards to all parts of the plant to show the path of water.
2. You should move your finger from the leaf to all the other parts of the plant to show the path of food.

Pages 124-125
A

1. The plant without leaves had not grown as well as the one with leaves. It was shorter.
2. The leaves need light to make them turn green.
3. The plant had grown to try to find light.

B

1. That too much or too little water can kill plants and there is a certain amount of water at which plants grow best (which is around 15cm3 per day for cress seedlings).
2. Warmth makes plants grow faster.

Page 126
A

1. Stone and brick 2. Wood, plastic and pottery.
3. Metal, rubber, glass. 4. Wooden table, fabric or plastic tablecloth, pottery or plastic plate, metal knife and fork

Page 127
A

1. Hard, rough, dull 2. Glass 3. A, D, E and F

Page 128
A

1. B has the most wear and A has the least.
2. C, D, A, B

Page 129
A

1. A and C 2. B, D and E 3. No
4. B lets through most water because there is a big spot on the towel. E and D have smaller spots.

Page 130
A

basalt – F, chalk – E, granite – B, slate – G, sandstone – A, marble – C, limestone – D

Page 131

A

1. Granite - non-porous, limestone - porous, sandstone - porous, basalt - non-porous
2. Non-porous because it keeps the rain from coming through the roof.

Page 132

A

1. Sand 2. Clay 3. By using a different sieve that only lets clay through and keeps silt in the sieve.

Page 133

B

1. A 40cm3, B 57cm3, C 29cm3 2. C 3. B

Pages 134-135

A

1. They repel each other. 2. They attract other. 3. They repel each other.
4. Attract 5. Repel

B

1. C 2. A 3. C

Page 136

A

1. A will squash, B will not change. 2. A will stretch a little, B will stretch a little.

Page 137

A

1. about 5cm 2. about 30cm

Page 138

A

Torch A

Page 139

A

Clear plastic T, window glass T, wood O, water T, frosted glass TR, brick O, cardboard O, metal O, greaseproof paper TR, orange juice O

Page 140

A

A 13.00 B 09.00 C 10.00

Page 141

A

1. 15cm 2. A - false, B - true, C - true

Page 142

A

Shin bone, kneecap, thigh bone, pelvis, spine

Page 143

A

slug - water skeleton
frog and snake - skeleton of bone
shrimp and spider - skeleton of armour

Page 144

The biceps gets shorter, harder and fatter.

Page 145

A

Makes the bones, joints, muscles and heart strong, prevents people becoming overweight.

Page 146

A

A is a habitat, B is a mini habitat, C and D are micro habitats.

Page 147

A

1. It would die because there is not enough light for it to survive.
2. It would be dashed on the rocks by the waves or be swept out to sea and die.

Answers

Page 148
A

1. Insects – D, spiders – A, fish – G,
amphibians – E, reptiles – B, birds – F
mammals – C
2. Humans belong to the mammal group.

Page 149
A

A – Horse chestnut, B – beech, C – ash,
D – holly, E – oak, F – lime

Page 150
A

1. lettuce (or any other plant food) > human
2. grain (or grass) > chicken (or sheep, cattle) >
human

Pages 152-153
A
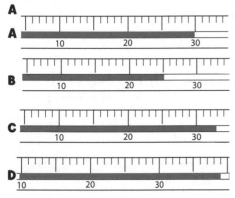

B
1.

Time from start (minutes)	Temperature (°C)
0	35
1	30
2	25
3	20
4	20

2. Because the water has cooled to room
temperature.

3.
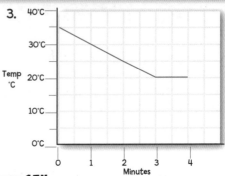

Page 154
A

1. The cup shows how the water cools without the
material. This makes the test fair by showing how
each material affects cooling.
2. B – wool
3. C – aluminium foil

Page 155
A

1. C – the steel spoon
2. The butter would melt because aluminium is a
metal like steel and metals are good conductors.
3. The heat passes quickly from the oven to the
food and speeds up cooking.
Pan handles are made of insulating materials like
wood or plastic so hands are not burned.

Page 156
A

1. A – 10cm3, B – 30cm3 2. 40cm3

Page 157
A

1. The same volume of liquid is used and the slope
is kept at the same angle.
2. A – vegetable oil, B – treacle, C – water

Page 158
A

1. A, C, B, D
2. B and D

3. 650°C

Page 159
B

The molten rock flowing down the side of a
volcano is a liquid and may be as hot as 1200°C.
It cools as it moves away and when it reaches
600°C it stops flowing and becomes solid rock.
The freezing point of the rock is 600°C.

Page 160
A

1. Sugar, instant coffee granules, salt
2. Some substances in the tea leaves dissolve in
the water but the leaf does not.

Page 161
A

1. The sand grains will stay in the filter paper
because they are too big to go through the
holes. The sugar will pass through because it has
dissolved in the water.
2. You can tell when sugar and salt are dissolved
in water by the taste of the water.
3. Yes. The holes let the dissolved substance pass
through but keep the leaves in the bag.

Page 162
A

A – 3, B – 5, C – 8, D – 4

Page 163
A

Sandpaper, velvet, newspaper, plastic sheet.

Page 164
A

1. Fastest – A, slowest – B
2. By repeating the experiment up to five times
more.

Page 165
A

1. C 2. B 3. 7 seconds

Page 166
A

motor – provides movement
wire – conducts electricity
buzzer – makes a sound
switch – controls the flow of electricity

Page 167
A

No. There is a gap in the circuit.

Page 168
A

1. steel spoon – conductor, wooden spoon –
insulator, copper pipe – conductor, plastic comb
– insulator
2. It is an insulator. If it was a conductor the
lamp would light when there was no other
material in the gap.

Page 169
A

When the tube is tipped so the left side goes
down, the ball bearing rolls onto the ends of the
wires and electricity can flow through it from one
wire to the next.

Page 170
A

A and D should be ticked.

Page 171
A

B

Notes

Notes

Notes

Index

Index